COMMIT TO SOLAR

The Ultimate Guide To Understanding, Marketing And Selling Solar

RUSS WARD

COPYRIGHTS

Commit To Solar: The Ultimate Guide To Understanding, Marketing And Selling Solar

All Rights Reserved.

ISBN: 978-1-329-00224-1
Copyright (c) 2023 Russ Ward

All rights reserved. No part of this book may be reproduced or transmitted in any form or by any means, electronic, or technical, including photocopying or recording, or by any information storage and retrieval system, without permission in writing by the author.

Table of Contents

Introduction 1

Benefits of Solar Energy 9

Marketing Solar Energy 61

Selling Solar Energy 109

Conclusion 148

Introduction

Thesolarthinking.net

Solar energy is more than just an environmentally friendly energy source. It can also help homeowners save money on their energy bills and provide a profitable opportunity for solar sales and marketing organizations. With the increasing demand for clean and sustainable energy, more households and businesses are turning to solar energy to reduce their energy costs and environmental impact.

In this book, we will explore the financial benefits of solar energy and provide strategies for homeowners to save money on their energy bills. We'll also provide tips for solar sales and marketing organizations to build a profitable business.

For homeowners, we'll show you how solar energy can reduce energy bills, increase home value, and provide a high return on investment. We'll also provide information on how to choose the right solar system for your home and how to maximize your savings.

For solar sales and marketing organizations, we'll cover the key strategies for selling solar to homeowners. We'll show you how to conduct a solar assessment, create a customized solar proposal, and effectively communicate the financial benefits of solar to homeowners. We'll also provide tips on how to overcome common objections and how to close more solar sales.

Marketing solar to potential customers can be a challenge, but we're here to help. We'll explore the key strategies for marketing solar to homeowners, including how to create effective marketing campaigns, leverage social media and other digital channels, and build a strong brand identity. We'll also provide tips on how to generate leads and how to effectively communicate with potential solar customers.

Overall, this book is your one-stop-shop for harnessing the financial benefits of solar energy. Whether you're a homeowner looking to save money on your energy bills or a solar sales and marketing organization looking to build a profitable business, this guide has the knowledge and tools you need to succeed.

Overview Of The Solar Energy Industry And Its Growth

Solar energy is energy obtained from the sun's radiation. It is a renewable and clean source of energy that has the potential to address many of the challenges associated with traditional energy sources such as fossil fuels.

Solar energy is collected through the use of solar panels that convert the sun's radiation into electricity or heat. There are two main types of solar panels: photovoltaic (PV) panels and solar thermal collectors. PV panels convert sunlight directly into electricity, while solar thermal collectors use the sun's energy to heat a fluid, which is then used to generate electricity or provide heat.

One of the primary reasons for the growing importance of solar energy is its environmental benefits. Solar energy does not produce greenhouse gas emissions, which are the primary contributors to climate change. This makes solar energy a clean and sustainable alternative to fossil fuels, which are responsible for a significant portion of global emissions.

In addition to its environmental benefits, solar energy is becoming increasingly important because of its economic benefits. The cost of solar panels has decreased significantly in recent years, making solar energy more affordable and accessible to households and businesses. In many cases, solar energy can now compete with traditional energy sources in terms of cost.

Finally, solar energy is important because it can help to address issues related to energy security. By generating electricity locally, households and businesses can reduce their dependence on centralized power grids, which are vulnerable to power outages and other disruptions.

Overall, the growing importance of solar energy is due to its potential to address some of the most pressing challenges of our time, including climate change, energy security, and economic development. As solar technology continues to improve and become more affordable, it is likely that solar energy will play an increasingly important role in meeting our energy needs in the future.

Explanation Of The Benefits Of Solar Energy And The Opportunities For Selling And Marketing It

Solar energy is a smart and attractive option for households and businesses for a variety of reasons. First and foremost, solar energy is a renewable and sustainable source of energy that will never run out. Unlike finite fossil fuels, which are becoming more expensive and difficult to obtain, solar energy is generated from an abundant and inexhaustible source – the sun.

Switching to solar energy can also help households and businesses save on their energy bills. By installing solar panels, they can generate their own electricity, reducing their reliance on the traditional power grid and lowering their energy bills.

The environmental benefits of solar energy are another significant advantage. Solar energy is a clean and sustainable source of energy that produces no emissions or pollution. By using solar energy, households and businesses can reduce their carbon footprint and help to mitigate climate change.

In addition to these benefits, solar energy can also provide energy independence for households and businesses. By generating their own electricity, they can reduce their dependence on centralized power grids, which are vulnerable to power outages and other disruptions.

Installing solar panels is also a long-term investment that can provide significant returns over time. Solar panels have a lifespan of up to 40 years and require minimal maintenance, making them a cost-effective and reliable source of energy.

The growing demand for renewable energy solutions presents significant opportunities for businesses to sell and market solar energy. Some potential marketing opportunities include highlighting the potential cost savings and long-term return on investment, emphasizing the environmental benefits, positioning solar energy as a way to achieve energy independence and resilience, educating customers about the benefits of solar energy and different installation and financing options, and offering strong customer support and warranties to ensure customer satisfaction and build brand loyalty.

Overall, solar energy provides a wide range of benefits, from economic savings and environmental benefits to energy independence and long-term investment potential. As demand for renewable energy continues to grow, businesses that market and sell solar energy can position themselves as leaders in the renewable energy sector and help create a more sustainable future for all.

Purpose Of This Book

The purpose of this book is to provide a comprehensive guide to solar energy for homeowners and solar sales and marketing organizations. The book will focus on three main areas: the benefits of solar, solar marketing, and solar sales.

The first section of the book will explore the benefits of solar energy and explain exactly how it works. This section will cover topics such as how solar energy can reduce energy bills, increase home value, and reduce carbon emissions. It will conclude by discussing the barriers to widespread adoption.

The second section of the book will cover solar marketing for organizations. This section will explore the key strategies for marketing solar to homeowners, including how to create effective marketing campaigns, how to leverage social media and other digital channels, and how to build a strong brand identity. It will also provide tips on how to generate leads and how to effectively communicate with potential solar customers.

The third section of the book will focus on solar sales for organizations. This section will cover the key strategies for selling solar to homeowners, including how to conduct a solar assessment, how to create a customized solar proposal, and how to effectively communicate the benefits of solar to homeowners. It will also provide tips on how to overcome common objections and how to close more solar sales.

Overall, the goal of this book is to provide a comprehensive guide to solar energy that is tailored to the needs of homeowners and solar sales and marketing organizations. By exploring the benefits of solar, the key strategies for selling solar, and the best practices for solar marketing, this book will provide readers with the knowledge and tools they need to effectively harness the power of solar energy.

Benefits of Solar Energy

1. Environmental Benefits

Solar energy is not just a source of renewable energy, it also offers a wide range of environmental benefits. One of the most obvious benefits is the reduction in carbon emissions. Unlike fossil fuel-based power plants that emit harmful pollutants, solar energy is a clean and sustainable source of energy that produces no emissions or pollution. By using solar energy, households and businesses can reduce their carbon footprint and help to mitigate climate change.

Another environmental benefit of solar energy is improved air quality. Fossil fuel-based power plants emit harmful pollutants that can contribute to respiratory illnesses and other health problems. Solar energy systems produce no air pollutants, which helps to improve air quality and protect public health.

Solar energy also helps conserve natural resources. Fossil fuel-based power plants require the extraction and processing of non-renewable resources such as coal and natural gas. In contrast, solar energy is generated from an abundant and renewable resource - the sun.

Furthermore, solar energy systems require no water for operation, which helps to conserve this precious resource. Fossil fuel-based power plants, on the other hand, require a significant amount of water for cooling and other purposes.

Solar energy can also help to reduce the amount of land required for energy production. Fossil fuel-based power plants often require large amounts of land for mining and extraction activities. Solar energy systems, on the other hand, can be installed on rooftops, parking lots, and other underutilized spaces, which helps to reduce the amount of land required for energy production.

Additionally, the use of solar energy can help to reduce the environmental impact of energy production and distribution.

By generating energy at the point of consumption, solar energy reduces the need for expensive infrastructure, such as new transmission lines and power plants. This helps to reduce the overall environmental impact of energy generation and distribution.

Finally, the use of solar energy can also help to protect wildlife and natural habitats. Fossil fuel-based power plants and transmission lines can disrupt wildlife habitats and migration patterns, and cause other environmental damage. Solar energy systems, on the other hand, have a much smaller environmental footprint and can be installed in ways that minimize the impact on wildlife and natural habitats.

Overall, solar energy provides a sustainable and environmentally friendly alternative to traditional fossil fuel-based energy sources. By harnessing the power of the sun, households and businesses can reduce their impact on the environment and help to create a more sustainable future. From reducing carbon emissions and air pollution to conserving natural resources and protecting wildlife habitats, solar energy is becoming an increasingly important tool for creating a healthier and more sustainable planet.

Reduction Of Greenhouse Gas Emissions

Solar energy is an essential component in the fight against climate change. The burning of fossil fuels for energy production is one of the leading contributors to greenhouse gas emissions, which have devastating impacts on the environment, including global warming, rising sea levels, and more frequent and severe natural disasters.

Solar energy provides a clean and sustainable alternative to traditional fossil fuel-based energy sources. By generating electricity from the sun, solar energy produces no emissions or pollution, significantly reducing greenhouse gas emissions and helping to mitigate the impacts of climate change.

The potential for solar energy to reduce greenhouse gas emissions is substantial. According to the International Energy Agency, the deployment of solar energy systems could help to reduce CO_2 emissions by up to 6 billion tons per year by 2050. This represents a significant reduction in greenhouse gas emissions and a major step towards a more sustainable future.

Moreover, the adoption of solar energy is critical in meeting the goals of the Paris Climate Agreement, which aims to limit global temperature increases to below 2 degrees Celsius. As countries around the world work to transition to clean energy sources, solar energy is emerging as a leading solution to reduce greenhouse gas emissions and combat climate change.

By investing in solar energy, individuals and businesses can play an active role in reducing greenhouse gas emissions and combating climate change. Whether it's installing solar panels on homes and businesses, investing in solar energy projects, or advocating for policies that support the growth of renewable energy, there are numerous ways to support the transition to a more sustainable future.

In addition to its environmental benefits, solar energy also provides a range of economic and social benefits, including reduced energy costs, improved energy security, and increased access to clean and reliable energy. As the world continues to embrace solar energy, it's clear that it is not only a vital tool in the fight against climate change but also an important driver of economic and social progress.

Decreased Dependence On Fossil Fuels

Solar energy is a critical tool for reducing our dependence on traditional fossil fuel-based energy sources. Fossil fuels, such as coal, oil, and natural gas, are finite resources that are rapidly depleting. As we continue to rely on these non-renewable resources, we face significant challenges, including rising energy costs, environmental degradation, and geopolitical tensions.

By harnessing the power of the sun, solar energy provides a clean, sustainable, and abundant energy source that reduces our dependence on fossil fuels. Solar energy systems generate electricity without burning fossil fuels or producing emissions, helping to mitigate the environmental impacts of traditional energy sources and provide a more sustainable energy future.

Reducing our dependence on fossil fuels also has economic benefits. Fossil fuels are subject to price volatility, and as they become more scarce, their cost is likely to rise. Solar energy, on the other hand, has a predictable and declining cost trajectory, making it an attractive option for businesses and homeowners looking to reduce their energy costs.

Moreover, the adoption of solar energy can help to improve energy security by reducing our reliance on centralized power grids, which are vulnerable to power outages and disruptions. Solar energy provides a decentralized energy source, allowing businesses and individuals to generate their own electricity and reduce their dependence on traditional energy providers.

By reducing our dependence on fossil fuels, we also reduce our environmental impact. Fossil fuels are responsible for a significant amount of greenhouse gas emissions, which contribute to climate change and other environmental problems. By embracing solar energy, we can reduce our carbon footprint and help to mitigate the impacts of climate change.

Overall, the adoption of solar energy provides an important opportunity to reduce our dependence on traditional fossil fuel-based energy sources. By embracing solar energy, we can improve our energy security, reduce our environmental impact, and provide a more sustainable energy future. As we continue to face the challenges of a changing world, solar energy provides a vital solution to meet our energy needs and support our economic, social, and environmental well-being.

Protection Of Wildlife And Habitats

The impact of traditional energy sources on wildlife and habitats is a growing concern. Fossil fuel extraction and transportation can have devastating effects on local ecosystems, including habitat destruction, water and air pollution, and the displacement or loss of wildlife populations.

Solar energy provides a sustainable and eco-friendly alternative to traditional energy sources that can help protect wildlife and their habitats. Unlike fossil fuels, solar energy generation requires no extraction or transportation of resources, which can have a significant impact on the environment.

Solar energy systems can be installed in a variety of locations, including on rooftops, parking lots, and other underutilized spaces. This reduces the need for large areas of land, which can help to preserve wildlife habitats and reduce the impact of energy production on local ecosystems.

In addition, solar energy systems produce no emissions or pollution, which can help to protect air and water quality and reduce the risk of negative impacts on wildlife populations. This can help support the health and well-being of both wildlife and humans in the surrounding areas.

The benefits of solar energy for wildlife and habitats extend beyond just the direct impact of energy production.

The adoption of renewable energy sources can also help to promote awareness and education about the need to protect natural resources and wildlife habitats. By embracing sustainable energy sources like solar, we can support a more responsible and eco-friendly approach to energy production and consumption.

Overall, solar energy provides an important opportunity to protect wildlife and their habitats from the negative impacts of traditional energy sources. By embracing solar energy, we can support the health and well-being of both wildlife and humans and create a more sustainable and responsible approach to energy production.

Cleaner Air And Water

One of the most significant benefits of solar energy is the reduction of harmful pollutants in the air and water. Traditional energy sources like coal and natural gas release a variety of harmful pollutants and greenhouse gases into the air, which contribute to climate change and can have serious health impacts on humans and wildlife alike.

Solar energy, on the other hand, generates electricity without releasing any harmful emissions or pollutants. This means that solar energy helps to keep the air and water cleaner and protect the health of both people and the environment.

In addition to directly reducing pollution, solar energy can also help to indirectly promote cleaner air and water. By reducing our reliance on fossil fuels and promoting sustainable energy sources, we can reduce the negative impacts of energy production on the environment and protect the quality of our air and water.

In particular, solar energy can help to reduce the amount of water needed for energy production. Fossil fuel-based power plants often require significant amounts of water for cooling and other purposes, which can strain local water resources and have negative impacts on water quality. Solar energy, on the other hand, requires no water for operation, which helps to conserve this precious resource.

By reducing pollution and promoting cleaner air and water, solar energy can have a significant positive impact on the environment and public health. The adoption of solar energy is an important step towards creating a more sustainable and responsible approach to energy production and consumption.

Overall, the benefits of solar energy for air and water quality are clear. By promoting solar energy and reducing our reliance on traditional energy sources, we can protect the environment, promote public health, and support a more sustainable future for generations to come.

2. Economic Benefits

Solar energy has become an increasingly attractive option for those looking to save money and build a more sustainable future. One of the biggest economic benefits of solar energy is the potential for reduced energy costs. By installing a solar energy system, households and businesses can generate their own electricity, reducing their reliance on the grid and saving money on energy bills. Over time, the cost savings from solar energy can be significant and can help to protect against rising energy costs.

Another economic benefit of solar energy is the potential increase in home value. Studies have shown that homes with solar energy systems installed have higher resale values compared to homes without solar. This means that homeowners can recoup some or all of their initial investment in a solar system when they sell their home.

The growing demand for solar energy has also created new job opportunities in manufacturing, installation, and maintenance of solar systems. This has helped to create new employment opportunities and support local economies. In fact, the solar industry is one of the fastest-growing job markets in the United States, with employment in the industry increasing by more than 150% in the past decade.

By using solar energy, households and businesses can also reduce their reliance on fossil fuel-based energy sources. This can help to create a more stable and sustainable energy supply and reduce the economic risks associated with fossil fuel price volatility and supply disruptions. For businesses, solar energy can also provide a competitive advantage by reducing operating costs and improving the bottom line.

Moreover, many governments and utilities offer financial incentives, such as tax credits and rebates, to encourage the adoption of solar energy. These incentives can help to reduce the cost of installing a solar energy system and improve the financial viability of solar projects. Some utilities even offer net metering, allowing customers to receive credits for excess energy generated by their solar system that is fed back into the grid.

Finally, solar energy can also help to reduce the overall cost of energy generation and transmission. By generating energy at the point of consumption, solar energy can reduce the need for expensive infrastructure, such as new transmission lines and power plants, helping to keep energy costs down for everyone.

Overall, solar energy provides a range of economic benefits that can help households, businesses, and society as a whole. By reducing energy costs, increasing home value, creating new job opportunities, reducing reliance on fossil fuels, offering government incentives, and reducing the overall cost of energy generation and transmission, solar energy is becoming an increasingly attractive option for those looking to save money and build a more sustainable future.

Lower Energy Costs

One of the most compelling economic benefits of solar energy is the potential for lower energy costs. With traditional energy sources, the cost of energy is subject to fluctuating fuel prices and supply chain disruptions, which can lead to unpredictable and expensive bills for households and businesses.

By contrast, solar energy provides a stable and predictable source of energy. By generating your own electricity, you can reduce your dependence on the traditional power grid and lower your energy bills. Once your solar panels are installed, the energy they generate is essentially free, as you are no longer relying on expensive fuel sources like coal or natural gas.

In fact, many homeowners and businesses report significant savings on their energy bills after installing solar panels. In some cases, solar energy can even help to offset the cost of the initial investment in solar panel installation. Additionally, many governments and utilities offer financial incentives, such as tax credits and rebates, to encourage the adoption of solar energy, which can further reduce the cost of installation.

The long-term financial benefits of solar energy are also significant. Solar panels have a lifespan of up to 40 years and require minimal maintenance, making them a cost-effective and reliable source of energy. In addition, by investing in solar energy, households and businesses can protect themselves against rising energy costs in the future.

The economic benefits of solar energy are not just limited to individual households and businesses. Solar energy also has the potential to create new job opportunities and support economic growth. As the demand for solar energy continues to grow, more jobs will be created in areas such as installation, maintenance, and research and development.

In summary, the economic benefits of solar energy are significant and wide-ranging. By reducing energy costs and creating new job opportunities, solar energy is becoming an increasingly important tool for promoting economic growth and sustainability.

Job Creation In The Solar Industry

The solar energy industry has been rapidly growing over the past few years, and with that growth has come a significant increase in job opportunities. As the demand for solar energy continues to rise, so does the need for skilled workers to install, maintain, and service solar energy systems.

One of the key economic benefits of solar energy is the job creation it provides. The solar industry employs a diverse range of workers, from engineers and electricians to sales representatives and project managers. In fact, the solar industry is one of the fastest-growing industries in the United States, with the number of solar jobs increasing by over 150% in the last decade.

The job opportunities in the solar industry are not limited to installation and maintenance roles. As the industry continues to evolve and mature, there is a growing need for professionals in marketing, finance, and legal positions. This means that there are opportunities for workers with a wide range of skills and backgrounds to enter the solar industry and make a positive impact on the world.

Furthermore, the job opportunities in the solar industry are not limited to one region or country. Solar energy is a global industry, and job opportunities can be found in countries all around the world. This means that workers in the solar industry have the potential to make a positive impact on a global scale.

For job seekers, the solar industry can provide a unique opportunity to not only make a positive impact on the environment, but also to build a fulfilling and rewarding career. The industry is constantly evolving and changing, which means that there are opportunities for workers to learn new skills and stay on the cutting edge of solar technology.

For businesses in the solar industry, job creation can also be a key selling point. By highlighting the job opportunities and positive impact on the local economy, businesses can differentiate themselves from competitors and build a strong reputation as a socially responsible organization.

Overall, the job creation in the solar industry is a significant economic benefit of solar energy. As the industry continues to grow and evolve, there will be even more opportunities for workers to build fulfilling careers and make a positive impact on the world.

Increased Property Values

Solar energy is increasingly becoming a popular option for homeowners and businesses. One of the key benefits of installing solar panels is the potential to increase the value of a property. This is because solar panels are seen as an asset that can generate income by selling excess energy back to the grid. Additionally, solar panels can significantly reduce energy bills for homeowners and businesses, making the property more attractive to potential buyers who are interested in long-term energy savings.

Another factor that can increase property values is the availability of tax credits and other incentives for installing solar panels. These incentives can further increase the value of a property and make it more attractive to potential buyers who are interested in the long-term financial benefits of owning a solar-powered property.

In addition to the financial benefits, the environmental benefits of solar energy can also increase the value of a property. Reduced carbon emissions, cleaner air and water, and other environmental benefits can be appealing to potential buyers who are environmentally conscious.

As more homeowners and businesses adopt solar energy, those without solar panels may be at a competitive disadvantage when it comes to property values. By installing solar panels, property owners can remain competitive and attract potential buyers who are interested in sustainable, energy-efficient properties.

Overall, the installation of solar panels can increase the value of a property in various ways, including cost savings, tax incentives, environmental benefits, and competitive advantage. These benefits make solar energy an attractive investment for homeowners and businesses looking to make a long-term investment in their property.

Stimulation Of Local Economies

The solar industry is not only a promising source of renewable energy but also a catalyst for economic growth, as it can stimulate local economies and create job opportunities in various fields.

One of the primary ways that the solar industry stimulates local economies is through job creation. Solar energy is labor-intensive and requires workers for various tasks, such as installation, maintenance, and research. As a result, the industry is expected to create many job opportunities, especially in areas where solar energy adoption is high. This can help to reduce unemployment rates and stimulate economic growth in local communities.

In addition, the growth of the solar industry can lead to increased tax revenues for municipalities and states. This is because the installation and maintenance of solar energy systems generate revenue for local businesses, which in turn contributes to local tax bases. Furthermore, the growth of the industry can attract new businesses and investors to the area, further contributing to economic growth.

Another way that the solar industry contributes to the economy is by reducing energy costs for households and businesses. By generating their own electricity through solar panels, households and businesses can reduce their reliance on the traditional power grid, which can result in significant cost savings over time. This can free up more money for local businesses and households to spend on other goods and services, further contributing to local economic growth.

Finally, the solar industry can promote sustainable growth and a cleaner environment. By reducing the dependence on fossil fuels, the solar industry can reduce carbon emissions, which can contribute to climate change. This can help protect public health and the environment and create a more sustainable future for generations to come.

Overall, the solar industry has significant potential to stimulate local economies and create long-term, sustainable growth. By creating job opportunities, generating tax revenues, reducing energy costs, and promoting a cleaner environment, the industry can contribute to economic growth and help build a brighter future for communities around the world.

3. Social Benefits

In addition to the environmental and economic benefits, solar energy provides a range of social benefits that can help to improve the quality of life for communities around the world.

One of the most important social benefits of solar energy is improving energy access for communities that are not connected to the grid or have limited access to reliable electricity. This can improve the quality of life and support economic development in these communities, providing opportunities for education, health care, and other vital services.

Solar energy can also improve energy security for households and businesses, particularly in areas that are prone to power outages. By generating their own electricity, they can reduce their reliance on the grid and ensure that critical infrastructure, such as hospitals and emergency services, can continue to operate during power outages.

The use of solar energy can also help to improve health and safety by reducing exposure to harmful pollutants and reducing the risk of power outages. Solar energy systems produce no emissions or pollution, which can help to improve air quality and protect public health.

Furthermore, the adoption of solar energy can help to raise awareness and educate communities about the benefits of renewable energy. This can help to support a more sustainable energy future and promote environmentally responsible behavior.

Solar energy can also help promote social justice by reducing energy poverty and supporting economic development in underserved communities.

By providing access to clean, reliable, and affordable energy, solar energy can help to reduce inequality and support social and economic development.

Finally, solar energy can help to create new job opportunities and support local economies. The growth of the solar industry has created new employment opportunities in manufacturing, installation, and maintenance of solar systems, helping to support local communities and promote economic development.

Overall, solar energy provides a range of social benefits that can help to improve the quality of life for communities around the world. By improving energy access and security, promoting health and safety, supporting education and awareness, promoting social justice, and creating new job opportunities, solar energy is becoming an increasingly important tool for creating a more sustainable and equitable future.

Improved Public Health

One of the great social benefits of solar is improved public health. Solar energy is a clean and renewable source of energy that produces no harmful emissions, unlike traditional fossil fuels, which release harmful pollutants into the air and water.

By reducing the amount of air pollution, solar energy can help to improve public health. Fossil fuel-based power plants emit harmful pollutants such as sulfur dioxide, nitrogen oxides, and particulate matter, which can contribute to respiratory illnesses, heart disease, and other health problems.

Solar energy systems produce no air pollutants, which helps to improve air quality and protect public health.

In addition, solar energy systems do not require large amounts of water for cooling and other purposes, unlike traditional power plants. This helps to conserve water resources and protect the environment.

Solar energy can also provide electricity to remote or underserved areas, improving access to reliable and sustainable energy sources. This can have a significant impact on the quality of life for people in these communities, as they no longer have to rely on costly and unreliable energy sources.

Furthermore, the adoption of solar energy can create job opportunities in local communities, which can have a positive economic impact. The growth of the solar industry creates jobs in manufacturing, installation, maintenance, and research, among other areas. This can help to stimulate local economies and provide opportunities for people in the community to earn a living.

Overall, solar energy provides several social benefits, including improved public health, access to sustainable energy sources, and job creation. By adopting solar energy, individuals and communities can improve their quality of life and promote a more sustainable future for themselves and future generations.

Increased Energy Independence and Security

With traditional energy sources, households and businesses are dependent on centralized power grids that are vulnerable to power outages and disruptions. However, by generating their own electricity through solar panels, households and businesses can become more energy independent and resilient. This can provide peace of mind and security in the event of power outages, natural disasters, or other disruptions.

Furthermore, solar energy can provide increased energy security for communities and nations as a whole. By reducing reliance on fossil fuels and foreign sources of energy, solar energy can enhance national security and reduce the potential for conflicts related to energy resources.

In addition to energy independence and security, solar energy can also improve social equity. Solar energy can provide access to electricity for individuals and communities who may not have access to traditional power grids, such as those in rural or remote areas.

This can improve the quality of life for individuals and increase economic opportunities for communities.

Moreover, solar energy can have a positive impact on public health. Traditional energy sources, such as coal and natural gas, emit pollutants that can harm human health and contribute to respiratory illnesses and other health problems. By reducing reliance on these sources and transitioning to clean, renewable energy sources like solar, public health can improve and medical costs can be reduced.

Overall, the social benefits of solar energy, including increased energy independence and security, improved social equity, and better public health, are significant and should not be overlooked. By promoting the adoption of solar energy and supporting policies that facilitate its expansion, we can work towards a more sustainable and equitable future.

Access To Electricity For Remote And Underserved Communities

Solar energy is an effective way to bring electricity to remote and underserved communities. Many of these communities lack access to reliable electricity due to their location or economic circumstances.

However, solar energy systems can provide a sustainable and affordable source of energy to power homes, schools, hospitals, and other essential services.

Solar energy can also help to reduce energy poverty, which is a major issue in many developing countries. By providing access to clean and sustainable energy, solar energy can improve the quality of life for individuals and families living in poverty. This can lead to improved health outcomes, better education, and increased economic opportunities.

In addition, solar energy can provide energy security to communities that may be vulnerable to power outages or disruptions. By generating their own electricity, these communities can reduce their dependence on centralized power grids, which are often unreliable in remote or underserved areas.

Overall, solar energy can have a significant social impact by providing access to electricity, reducing energy poverty, and improving energy security in remote and underserved communities. By promoting the benefits of solar energy and providing support for the development of solar infrastructure, we can help to create a more sustainable and equitable energy future for all.

Thesolarthinking.net

Empowerment Of Communities Through Community-owned Solar Projects

---

Solar energy provides a unique opportunity for communities to take ownership of their energy future through community-owned solar projects. By investing in solar energy, communities can generate their own clean and sustainable energy, reducing their reliance on traditional energy sources and improving their energy security.

One of the key social benefits of community-owned solar projects is community empowerment. By working together to develop and implement solar projects, communities can become more self-sufficient and take control of their energy needs. This can lead to increased community pride and a sense of accomplishment, as well as greater engagement in local decision-making.

Community-owned solar projects can also provide economic benefits to communities. By generating their own energy, communities can reduce their energy bills and save money over time. In addition, solar projects can create jobs in areas such as installation, maintenance, and administration, providing economic opportunities for local residents.

Another social benefit of community-owned solar projects is increased access to energy for underserved communities. In many cases, low-income and underserved communities have limited access to traditional energy sources. Community-owned solar projects can provide these communities with access to clean and sustainable energy, improving their quality of life and reducing their energy costs.

Overall, community-owned solar projects provide social benefits such as community empowerment, economic opportunities, and increased access to energy. By working together to develop and implement solar projects, communities can take control of their energy future and create a more sustainable and equitable energy system.

4. Technological Benefits

In addition to the environmental, economic, and social benefits, solar energy also provides a range of technological benefits that make it an attractive option for households and businesses.

One of the most important technological benefits of solar energy is that it is modular and scalable. Solar panels can be installed in a variety of sizes and configurations, making it possible to customize the energy generation capacity to meet the specific needs of a home or business.

This flexibility can also make solar energy a more affordable option, since homeowners and businesses can start with a small system and gradually add to it over time as their energy needs grow.

Moreover, solar energy systems require very little maintenance and can last for decades with minimal upkeep. This makes them a highly reliable source of energy, with very little downtime or interruption in service.

Solar energy systems also have no moving parts, which means they are virtually silent and do not produce vibrations that can cause wear and tear over time. This makes them an ideal choice for homes and businesses that require a quiet and reliable source of energy.

Furthermore, solar energy systems are highly efficient, converting a high percentage of the sun's energy into usable electricity. This efficiency has improved significantly over the past few decades, making solar energy a more cost-effective and viable option for households and businesses.

In addition, solar energy can be used in combination with other energy sources, such as wind or hydroelectric power, to create hybrid energy systems. This can help to provide a more stable and reliable source of energy, reducing the impact of weather and other external factors on energy generation.

Finally, advances in solar technology have made it possible to integrate solar energy systems into a range of building designs, from residential homes to commercial buildings. This means that solar panels can be integrated into the design of a building, making them a more aesthetically pleasing and visually appealing option.

Overall, solar energy provides a range of technological benefits that make it an attractive and reliable option for households and businesses. From its modular and scalable design to its low-maintenance and highly efficient operation, solar energy is becoming an increasingly important tool for creating a more sustainable and reliable energy future.

Advancements In Solar Technology

Solar technology has come a long way in recent years, with significant advancements being made in the development of solar panels and other related technologies. Some of the technological benefits of solar energy include:

Efficiency: With the latest advancements in solar technology, solar panels are becoming more efficient at converting sunlight into electricity. This means that smaller and more compact solar panels can generate more electricity than larger, less efficient ones.

Storage: One of the biggest technological challenges with solar energy has been storing excess energy for use when the sun is not shining. However, new battery technologies are making it possible to store solar energy for later use, making solar energy a more reliable source of electricity.

Smart Grids: Advances in solar technology are also leading to the development of smarter power grids. These grids can integrate renewable energy sources, like solar, into the traditional power grid, allowing for a more efficient and sustainable distribution of electricity.

Innovation: As solar technology continues to improve, new and innovative applications are being developed. For example, researchers are exploring the use of solar-powered transportation and solar-powered desalination plants to provide clean drinking water in areas with limited access to fresh water.

Overall, the technological benefits of solar energy are making it an increasingly viable and attractive option for households and businesses looking to reduce their carbon footprint and lower their energy costs. With ongoing innovation and improvements in solar technology, the future looks bright for the solar industry.

Increased Energy Efficiency

One of the main technological benefits of solar is increased energy efficiency. Solar panels can convert the energy from the sun into electricity with high efficiency rates, reducing the amount of energy lost during the conversion process. This means that households and businesses can generate more electricity from the same amount of sunlight, leading to higher energy output and reduced waste.

In addition to improved efficiency, solar technology has also led to the development of new energy storage solutions. With the use of batteries and other energy storage technologies, solar energy can be stored and used at night or during periods of low sunlight. This has led to increased energy independence and reduced reliance on traditional power grids.

Another technological benefit of solar energy is its versatility. Solar panels can be installed in a variety of locations, from rooftops and parking lots to open fields and deserts. This flexibility allows for solar energy to be used in a range of settings, including in remote and off-grid locations.

Finally, advancements in solar technology have also led to improvements in the design and aesthetics of solar panels. Panels can now be designed to blend in with the surrounding environment or even serve as building materials, such as roof tiles or facades. This has made solar energy a more attractive and viable option for households and businesses looking to incorporate renewable energy into their properties.

Overall, the technological benefits of solar energy have led to improved efficiency, energy storage solutions, versatility, and aesthetics. These advancements have made solar energy an increasingly attractive option for households, businesses, and communities looking to reduce their reliance on traditional energy sources and promote a more sustainable future.

Integration With Other Renewable Energy Sources

Solar energy is an important part of the renewable energy landscape, and one of its key technological benefits is its ability to integrate with other renewable energy sources. As the world seeks to reduce its reliance on fossil fuels and transition to more sustainable energy sources, solar energy can play a vital role in supporting this transition by working in tandem with other renewable energy technologies.

One of the most promising areas of integration is the pairing of solar energy with energy storage systems, such as batteries.

By storing excess solar energy during times of low demand and releasing it during peak demand periods, energy storage systems can help to address one of the biggest challenges of solar energy – its intermittent nature.

Another promising area of integration is the use of solar energy in combination with wind power. While solar energy is most abundant during the day, wind energy is typically more abundant at night. By combining these two energy sources, it is possible to create a more consistent supply of renewable energy throughout the day and night.

Solar energy can also be used in conjunction with other renewable energy sources, such as hydropower, geothermal energy, and biomass energy. By working together, these technologies can provide a more robust and reliable energy supply while reducing our reliance on fossil fuels.

Furthermore, solar energy is helping to drive technological advancements in other areas, such as energy storage, smart grid technology, and electric vehicle charging infrastructure. As solar energy continues to gain popularity, it is spurring innovation and investment in related technologies, which in turn is driving down costs and making renewable energy more accessible to a wider range of people and businesses.

In conclusion, the technological benefits of solar energy are numerous, and its ability to integrate with other renewable energy sources is a key factor in driving the transition to a more sustainable energy future. As solar technology continues to evolve and improve, it will become an even more important piece of the renewable energy puzzle, helping to reduce our reliance on fossil fuels and create a cleaner, more sustainable world.

Improved Energy Storage Solutions

One of the main benefits of the technological advances of solar is the development of improved energy storage solutions. With solar panels, energy is produced during the day when the sun is shining, but there is no energy produced at night when the sun has set.

This is where energy storage comes into play. By storing excess energy produced during the day, households and businesses can use the stored energy at night or during periods of low solar production, reducing their reliance on the traditional power grid and ensuring a more stable and reliable source of energy.

Another technological benefit of solar energy is its integration with other renewable energy sources. This allows for a more diverse and flexible energy mix, making it easier to meet the energy needs of households and businesses. For example, solar panels can be integrated with wind turbines, hydropower systems, and other renewable sources to provide a more stable and reliable source of energy.

In addition, solar energy has contributed to advancements in energy efficiency. Solar panels are now more efficient than ever, with higher conversion rates and increased durability. This means that more energy can be produced using fewer panels, making solar energy more cost-effective and practical for households and businesses.

Overall, the technological benefits of solar energy, including improved energy storage solutions, integration with other renewable energy sources, and increased energy efficiency, make it a promising source of energy for the future. As advancements in solar technology continue, it is likely that we will see even more benefits and opportunities for the use of solar energy in the years to come.

5. Barriers To Adoption

There are several barriers that have hindered the widespread adoption of solar energy in many places. One of the most significant challenges is the initial cost of installation. While the long-term cost savings can be significant, the upfront cost can be a major deterrent for many households and businesses.

Another barrier is the lack of suitable infrastructure to support solar energy. In many cases, the traditional power grid is not designed to accommodate the intermittent nature of solar energy production. This can lead to challenges with energy storage and distribution.

Additionally, there is a lack of widespread awareness and understanding of the benefits of solar energy. Many people are simply not familiar with how solar energy works or the potential cost savings and environmental benefits it can provide.

Finally, there can be regulatory and policy barriers that make it difficult to install solar panels. For example, certain areas may have zoning laws or homeowner association rules that restrict the installation of solar panels, or there may be issues with obtaining permits or meeting building codes.

Overall, these barriers can present significant challenges to the widespread adoption of solar energy. However, with continued advancements in technology, greater awareness of the benefits of solar energy, and supportive policies and regulations, it is possible to overcome these barriers and build a more sustainable energy future.

Cost And Financing Challenges

When it comes to the adoption of solar energy, one of the major barriers is the cost and financing challenges associated with installation.

The initial cost of installing solar panels can be high, making it difficult for some households and businesses to afford the upfront expense.

Furthermore, financing can be a challenge for those who cannot afford to pay for the installation outright. Although there are options such as leasing or power purchase agreements, these often come with their own complications and limitations.

Another challenge is the lack of understanding about the financial benefits of solar energy. Some people may not be aware of the long-term savings that can be achieved through reduced energy bills and potential income generated from selling excess energy back to the grid.

In addition, there may be regulatory and bureaucratic barriers to the installation of solar panels. Obtaining permits and navigating regulations can be a complex and time-consuming process, which may deter some from pursuing solar energy.

Finally, some people may be resistant to change and reluctant to adopt new technologies, particularly if they are not familiar with them or are unsure of their reliability.

Despite these barriers, there are several solutions and strategies that can be employed to overcome them. For example, public education and outreach efforts can help to increase awareness about the benefits of solar energy and dispel misconceptions. Additionally, financial incentives and tax credits can help to make solar energy more affordable and accessible.

Furthermore, simplifying the permitting and installation process can make it easier and more appealing for households and businesses to pursue solar energy. As technology continues to advance, the cost of solar installation is expected to decrease, making it more accessible to a wider range of people.

Overall, while there are challenges to the adoption of solar energy, there are also numerous solutions and strategies that can be employed to overcome them. By addressing these challenges and promoting the benefits of solar energy, we can work towards a more sustainable and energy-efficient future.

Lack Of Education And Awareness

Solar energy is a clean and renewable source of energy that offers numerous benefits to households and businesses. However, despite its advantages, the adoption of solar energy has been hindered by several barriers. One of the significant barriers to adoption is the lack of education and awareness about solar energy.

Many people are not aware of the benefits of solar energy, how it works, or the different options available for installation and financing. This lack of awareness can make potential customers skeptical about the effectiveness and cost-effectiveness of solar energy, which can discourage them from adopting it.

In addition, the lack of education about solar energy also affects the decision-making process for policymakers and regulators. Without sufficient knowledge about solar energy, policymakers and regulators may be hesitant to implement policies that promote the use of solar energy, such as tax credits and incentives.

To overcome this barrier, it is essential to educate and raise awareness about solar energy. This can be achieved through public education campaigns, informational materials, and workshops that provide potential customers with a better understanding of the benefits and options for solar energy adoption.

Overall, overcoming the barriers to the adoption of solar energy, such as lack of education and awareness and cost and financing challenges, requires a collaborative effort between policymakers, businesses, and communities. By working together to increase awareness, promote accessible financing, and offer incentives, we can create a more sustainable future powered by solar energy.

Policy And Regulatory Barriers

As solar energy gains popularity, there are still several barriers to its widespread adoption. One of the major challenges is policy and regulatory barriers. Some policies and regulations make it difficult or costly for households and businesses to install solar panels. For example, some utility companies have policies that make it harder for customers to sell excess solar energy back to the grid. Other policies and regulations may not provide adequate incentives for solar energy adoption, or may place restrictions on where solar panels can be installed.

Another barrier to adoption is the lack of education and awareness about solar energy. Many people are still unfamiliar with how solar energy works, its benefits, and the available financing options. This lack of awareness can make it difficult for solar energy companies to reach potential customers and promote the benefits of solar energy.

Cost and financing challenges are also a significant barrier to solar energy adoption. While the cost of solar panels has decreased over the years, it can still be a significant investment for households and businesses. Financing options, such as loans or leases, can make solar energy more accessible, but these options may not be available or suitable for everyone.

Finally, installation and maintenance challenges can also be a barrier to solar energy adoption. Some households and businesses may not have the necessary roof space or structural support to install solar panels. Maintenance and repairs can also be a challenge, as not all solar panel installers may have the necessary expertise or resources.

Despite these barriers, the benefits of solar energy, such as reduced energy costs and environmental sustainability, make it a promising energy source for the future. By addressing these barriers through education, policy changes, and innovative financing options, we can help make solar energy more accessible to all.

Net Metering and the Political Challenges

Net metering is a crucial policy for the solar industry that allows solar customers to sell excess energy back to the utility at retail rates. This policy is essential for incentivizing solar adoption and making solar energy more accessible and affordable for homeowners and businesses. Net metering has been a significant driver of the growth of the solar industry, but it also faces political challenges.

Net metering works by allowing solar customers to offset their energy bills by producing their electricity. When a solar system generates more energy than is consumed, the excess energy is sent back to the grid, and the customer receives a credit for the excess energy. This credit can then be used to offset future energy bills when the solar system is not generating enough energy.

The benefits of net metering are clear. It incentivizes solar adoption, reduces energy bills, and promotes energy independence. However, the implementation of net metering has faced opposition from some utilities and policymakers, who argue that net metering shifts the cost of maintaining the electrical grid to non-solar customers.

Utilities have pushed for changes to net metering policies, such as reducing the credit for excess energy, adding fixed fees for solar customers, and implementing demand charges. These changes make solar less attractive to potential customers and create financial challenges for solar businesses.

The political challenges surrounding net metering are complex. On the one hand, utilities and policymakers argue that net metering policies shift the cost of maintaining the grid to non-solar customers, and on the other hand, advocates for net metering argue that solar customers deserve fair compensation for the excess energy they produce and sell back to the grid.

There are also challenges associated with the implementation of net metering policies. Each state has its own net metering policy, which can create confusion and uncertainty for solar customers and businesses. The policies can also be subject to change, which can make it difficult for solar businesses to plan for the future.

To overcome the political challenges associated with net metering, it is essential to build coalitions and partnerships with stakeholders, including utilities, policymakers, and other industry players. Building relationships and working collaboratively can help to find common ground and develop policies that benefit everyone.

In conclusion, net metering is a crucial policy for the solar industry that has faced political challenges related to shifting the cost of maintaining the grid and implementation issues. Building coalitions and partnerships with stakeholders can help to overcome these challenges and develop policies that support the growth of the solar industry. By advocating for fair compensation for solar customers and promoting policies that incentivize solar adoption, we can create a more sustainable and equitable energy future for all.

How Energy Storage Can Help Overcome The Challenges Of Net Metering

---

As the demand for solar energy continues to grow, so too does the demand for more efficient and cost-effective energy storage solutions. Solar batteries are becoming an increasingly popular way to store excess solar energy generated during the day, for use when the sun is not shining.

One of the key benefits of solar batteries is that they can help overcome the challenges of net metering. With net metering, excess energy generated by a solar system is fed back into the grid, and the customer is credited for the excess energy produced. However, net metering policies vary by state, and some states are phasing out or reducing the benefits of net metering. This can make it difficult for solar customers to see a return on their investment, especially if they are generating more energy than they can use during peak sun hours.

By using solar batteries, customers can store their excess energy for later use, instead of feeding it back into the grid. This allows them to use their stored energy when they need it, rather than relying on the grid. It also means that they can continue to benefit from solar energy even if net metering policies change.

However, one of the main challenges to widespread adoption of solar batteries is the high cost. The cost of solar batteries must come down for this technology to be more widely adopted. As more companies invest in research and development, and as production increases, the cost of solar batteries is expected to decrease. In the meantime, some states and utilities are offering incentives and rebates to help offset the cost of solar batteries, making them a more affordable option for solar customers.

Technical Challenges

The adoption of solar energy is not without its challenges, particularly in terms of the technical aspects of implementing solar systems. One of the biggest technical challenges of solar energy is the intermittency of sunlight. Unlike traditional power sources that provide consistent energy, solar energy is dependent on the availability of sunlight. This can create challenges in providing consistent energy to households and businesses, particularly in areas with inconsistent weather patterns or limited sunlight.

Another technical challenge of solar energy is the issue of energy storage. Solar panels only produce energy during daylight hours, which means that excess energy must be stored for later use. While battery technology has improved significantly in recent years, energy storage remains a significant challenge for widespread solar adoption, particularly in areas where electricity grids are unreliable or non-existent.

The installation of solar panels can also be a technical challenge, particularly for older buildings or buildings with limited roof space. The weight of solar panels and the necessary mounting hardware can put stress on older roofs, requiring reinforcement or repair before installation can take place.

Finally, there is a need for specialized knowledge and expertise in the design, installation, and maintenance of solar systems. This requires a significant investment of time and resources for both households and businesses, and can be a barrier to adoption for those who lack the necessary knowledge or resources.

Overall, while solar energy offers numerous benefits, there are also significant technical challenges to widespread adoption. As technology continues to improve and the industry matures, these challenges may be overcome, making solar energy a more viable and accessible option for households and businesses.

The Solar King's Advice to Solar Sales And Marketing Professionals

In the field of solar energy sales, professionals are called upon to do more than simply sell a product or service - they must also fully embrace the mission and purpose behind it. Success in this industry requires a commitment to promoting sustainable and renewable energy solutions, and a passion for making a positive impact on the environment and the communities they serve.

To be successful at the highest levels:

- You must completely sell out to the concept of solar.

- You must be a subject matter expert on the benefits of solar and be able to explain them in a way everyone you meet can easily understand.

- You must actively take part in working toward a more sustainable, green energy future.
- If your home is eligible for solar panels you should strongly consider having them installed on your own home.

The Solar King's Additional Resources For Solar Marketers And Sales Organizations

- The Solar King's Free Solar Marketing Resources
  https://thesolarking.net/free-solar-resources
- The Solar King's Digital Marketing Course
  https://thesolarking.net/solar-leads-course
- The Solar King's Done For Your Marketing Package:
  https://thesolarking.net/offer
- The Solar King's Coaching Programs
  https://thesolarking.net/coaching
- The Solar King's Live Training Events
  https://thesolarking.net/events

Marketing Solar Energy

Thesolarthinking.net

1. Understanding Your Target Market

Understanding your target market is essential for developing effective marketing strategies for selling solar energy. By understanding the needs, priorities, and values of your target market, you can develop marketing messages and campaigns that resonate with potential customers and help to build brand loyalty. Here are some strategies for understanding your target market when marketing solar energy:

One of the first steps in understanding your target market is to conduct market research. Market research can help you to identify customer segments, understand their needs and priorities, and develop targeted marketing messages that resonate with those segments. Market research can be conducted through surveys, focus groups, and other methods, and can provide valuable insights into customer attitudes and behavior.

Another strategy for understanding your target market is to segment your customer base. By dividing customers into distinct segments based on their demographics, behavior, and other factors, businesses can develop targeted marketing messages and campaigns that are tailored to the specific needs and priorities of each segment. Common customer segments for solar energy include homeowners, businesses, and government organizations.

Once you have identified your target market and segmented your customer base, it is important to develop marketing messages that resonate with your customers. Marketing messages should emphasize the economic, environmental, and energy independence benefits of solar energy, and should be tailored to the needs and priorities of different customer segments.

For example, marketing messages for homeowners might emphasize the cost savings associated with switching to solar energy, while marketing messages for businesses might emphasize the energy independence and resilience benefits of solar energy.

Finally, it is important to develop a strong brand identity that resonates with your target market. A strong brand identity can help to build brand recognition and loyalty, and can help to differentiate your business from competitors. Brand identity should be based on the values and priorities of your target market, and should be communicated consistently across all marketing channels.

Overall, understanding your target market is essential for developing effective marketing strategies for selling solar energy. By conducting market research, segmenting your customer base, developing targeted marketing messages, and developing a strong brand identity, businesses can attract new customers and establish themselves as leaders in the renewable energy sector.

Demographic Analysis Of Potential Customers

As a solar energy business or marketing organization, understanding your target market is crucial to your success. A key part of this is conducting a demographic analysis of potential customers. This involves gathering information about your target market, including their age, gender, income, education level, and geographic location.

One important demographic factor to consider is age. For example, younger generations may be more interested in sustainable energy solutions and more willing to invest in solar energy. Additionally, income level is important to consider, as solar energy systems can be a significant upfront investment. Understanding the income level of potential customers can help you develop financing options to make solar energy more accessible to a wider range of customers.

Another important factor to consider is geographic location. Solar energy is more commonly adopted in areas with high levels of sun exposure and where electricity costs are high. Understanding the energy market in the region where you operate can help you develop marketing and pricing strategies that appeal to your target market.

Education level is also an important demographic factor to consider. Customers with a higher level of education may have a greater understanding of the environmental benefits of solar energy and be more likely to invest in solar systems. Therefore, it may be necessary to educate potential customers about the benefits of solar energy and how it can help them save money on energy bills.

By understanding the demographics of potential customers, you can develop targeted marketing strategies that appeal to their specific needs and interests. This can help you effectively reach your target market and increase the adoption of solar energy in your community.

It is important to note that demographic analysis is just one part of understanding your target market. Other factors, such as behavior and psychographics, also play an important role in developing effective marketing strategies. By taking a holistic approach to understanding your target market, you can develop strategies that effectively reach and engage potential customers, and increase the adoption of solar energy.

Identifying Customer Needs And Pain Points

As a solar energy marketer, it's crucial to understand the needs and pain points of potential customers. Conducting a demographic analysis can help you identify these needs and understand what motivates your target market to invest in solar energy.

To begin, you need to understand the demographic characteristics of your potential customers. This includes factors such as age, income, location, education, and more. These demographics can provide insights into what types of households or businesses are more likely to invest in solar energy.

Once you have identified the demographic characteristics of your target market, you need to analyze their needs and pain points. This includes understanding their current energy usage, the cost of their energy bills, and any issues they may be experiencing with their current energy provider. By understanding these pain points, you can tailor your marketing message to show how solar energy can solve their problems and provide a better solution.

For example, if your target market is households with high energy bills, you could tailor your message to highlight the potential cost savings of installing solar panels. You could emphasize how solar energy can reduce their monthly energy bills, helping them to save money in the long run. Alternatively, if your target market is businesses that are concerned about their carbon footprint, you could focus on the environmental benefits of solar energy, highlighting how it can help them to become more sustainable and responsible corporate citizens.

By conducting a demographic analysis and understanding the needs and pain points of your potential customers, you can create a targeted marketing message that resonates with your target market. This can help to increase your conversion rates and attract more customers to the benefits of solar energy.

Understanding Customer Buying Behavior

Understanding customer buying behavior is critical to the success of any marketing campaign, including those focused on selling solar energy. By understanding how customers make decisions and what factors influence those decisions, marketers can develop targeted campaigns that are more likely to resonate with potential customers.

One key aspect of customer buying behavior is understanding the customer journey. This involves understanding the various stages a customer goes through as they become aware of, consider, and ultimately decide to purchase a product or service. For solar energy, this journey might involve customers first becoming aware of the benefits of solar energy, researching different solar energy providers, comparing costs and benefits, and ultimately making a decision on whether to move forward with installation.

Another important aspect of customer buying behavior is understanding customer motivations and pain points. For solar energy, this might include understanding why a potential customer is interested in solar energy, such as a desire to reduce their carbon footprint, lower their energy costs, or increase the value of their property. It might also involve understanding any concerns or hesitations potential customers might have, such as worries about the initial cost of installation or doubts about the reliability of solar energy.

Demographic analysis can also play a role in understanding customer buying behavior. By examining the characteristics of potential customers, such as age, income level, and location, marketers can develop targeted campaigns that are more likely to appeal to specific groups. For example, marketing efforts might be tailored to appeal to young families who are concerned about the long-term financial benefits of solar energy, or to retirees who are interested in increasing the value of their property.

In summary, understanding customer buying behavior is critical to the success of marketing efforts focused on selling solar energy. By understanding the customer journey, identifying customer motivations and pain points, and conducting demographic analysis, marketers can develop targeted campaigns that are more likely to resonate with potential customers and lead to successful sales.

Competitor Analysis

One of the key aspects of developing a successful solar energy marketing strategy is understanding your competition. This means conducting a thorough analysis of your competitors to identify their strengths, weaknesses, and the unique value propositions they offer to potential customers.

To begin, it's important to identify who your direct competitors are within the solar energy industry. This includes not only other solar panel installation companies, but also other renewable energy providers and even traditional energy providers. Once you've identified your competitors, you can then conduct a thorough analysis of their business model, target market, and marketing strategies.

In addition to understanding your competitors, it's also important to understand your potential customers. Demographic analysis can help you identify the specific needs and pain points of your target audience, as well as their buying behaviors and preferences. By understanding your customers, you can better tailor your marketing strategy and product offerings to meet their needs and preferences.

It's also important to consider the unique value propositions that you can offer to potential customers. This may include unique financing options, innovative product offerings, or exceptional customer service. By identifying your unique strengths and value propositions, you can differentiate yourself from your competitors and attract potential customers who are looking for something that your competitors don't offer.

Overall, conducting a thorough analysis of your competitors and potential customers is essential for developing a successful solar energy marketing strategy. By understanding your competition and customers, you can identify gaps in the market and develop a unique value proposition that sets you apart from the competition.

2. Developing A Marketing Plan

Developing a marketing plan is essential for businesses that sell solar energy. The plan should help to identify target markets, develop marketing messages and campaigns, and allocate resources effectively.

To begin with, the target market must be defined. This includes identifying ideal customers and understanding their needs and priorities. By doing so, the marketing messages and campaigns can be tailored to the specific needs and priorities of the target market, which can help to build brand loyalty.

After defining the target market, it's important to develop marketing messages and campaigns that resonate with potential customers. These messages should emphasize the economic, environmental, and energy independence benefits of solar energy, and they should be communicated consistently across all marketing channels.

Next, businesses should identify the marketing channels that are most effective for reaching their target market. This may include online advertising, social media, email marketing, and events. By allocating resources effectively, businesses can ensure that their marketing efforts are focused on the most effective channels and campaigns.

Developing a budget is an essential part of any marketing plan. The budget helps to allocate resources effectively and ensures that marketing efforts are focused on the most effective channels and campaigns. Businesses should consider both short-term and long-term marketing goals when developing a budget.

Finally, it's important to evaluate the results of marketing efforts on an ongoing basis. By tracking key performance indicators (KPIs) such as website traffic, lead generation, and customer acquisition, businesses can adjust their marketing strategies as needed to achieve marketing goals.

In summary, developing a marketing plan is crucial for businesses that sell solar energy. By defining target markets, developing marketing messages and campaigns, identifying marketing channels, developing a budget, and evaluating results, businesses can attract new customers and establish themselves as leaders in the renewable energy sector.

Setting Marketing Goals And Objectives

When it comes to marketing solar energy, having a well-planned and executed marketing plan is essential to the success of the business. Developing a marketing plan involves several key steps, including setting marketing goals and objectives.

Marketing goals and objectives help to provide direction for the marketing plan and ensure that the efforts are aligned with the overall business objectives. It is important to set specific, measurable, achievable, relevant, and time-bound (SMART) goals and objectives that are aligned with the needs and desires of the target market.

One common marketing goal for solar energy businesses is to increase awareness and understanding of the benefits of solar energy among potential customers. This can be achieved through various marketing activities such as social media marketing, email marketing, and content marketing.

Another marketing goal may be to generate leads and increase the number of sales. This can be accomplished through targeted advertising and sales promotions, such as discounts or special offers for new customers.

To achieve marketing goals and objectives, it is important to identify and prioritize marketing strategies and tactics. This can involve conducting market research to better understand the needs and desires of potential customers, as well as analyzing the competition and their marketing strategies.

Once marketing goals, objectives, and strategies have been established, the next step is to develop a detailed action plan. This may involve creating a content calendar for social media, planning specific advertising campaigns, or setting up a system for tracking and following up on leads.

Regular monitoring and evaluation of the marketing plan is also important to ensure that the strategies and tactics are effective and achieving the desired results. Adjustments and revisions can be made as needed based on the analysis of data and feedback from customers.

In summary, developing a marketing plan for solar energy involves setting SMART goals and objectives, identifying target markets, analyzing competition, prioritizing marketing strategies, and developing a detailed action plan. Regular monitoring and evaluation of the marketing plan is also essential to ensure that it remains effective and aligned with the overall business objectives.

Defining Your Brand And Messaging

One of the key elements of a successful marketing plan is defining your brand and messaging.

Your brand is more than just a logo or a name. It represents the values, mission, and personality of your company. When it comes to marketing solar energy, it is important to establish a brand that is associated with sustainability, innovation, and reliability.

Once you have established your brand, you need to develop messaging that is consistent with it. Your messaging should clearly communicate the benefits of solar energy to your target audience in a way that resonates with them. This requires a deep understanding of your target market and the specific needs and pain points that they have.

Your messaging should be clear, concise, and compelling. It should highlight the benefits of solar energy, such as cost savings, environmental sustainability, and energy independence. It should also address any concerns or objections that potential customers may have, such as the upfront cost of installation or the reliability of solar energy.

In addition to defining your brand and messaging, your marketing plan should also outline the specific tactics and channels that you will use to reach your target audience. This may include digital marketing, traditional marketing, and community outreach.

Overall, developing a clear and consistent brand and messaging is essential for effective solar energy marketing. By establishing a strong brand and communicating the benefits of solar energy in a compelling way, you can attract and retain customers and build a reputation as a leader in the renewable energy industry.

Creating A Budget And Allocating Resources

One of the most important steps in creating a marketing plan is to set a budget and allocate resources. This includes determining how much money can be spent on marketing efforts, and deciding how to divide that budget among different marketing channels.

When creating a marketing budget, it is important to consider the costs associated with various marketing strategies, such as digital marketing, traditional marketing, events, and sponsorships. It is also important to allocate resources effectively to ensure that marketing efforts are targeted towards the most profitable customer segments.

Once the marketing budget has been established, resources can be allocated to specific marketing channels. This may involve hiring marketing professionals, investing in software and technology, and outsourcing marketing tasks to third-party providers.

In addition to setting a budget and allocating resources, it is important to create a clear and consistent brand message. This involves defining the unique selling proposition of the solar energy product, and communicating this message through all marketing channels. This helps to build brand awareness and establish the product as a leader in the market.

Overall, developing a marketing plan is essential for successfully promoting and selling solar energy. By setting marketing goals and objectives, defining the brand message, and creating a budget and allocating resources, businesses can create effective marketing campaigns that reach their target audience and drive sales.

Determining The Best Marketing Channels For Your Target Audience

When it comes to marketing solar energy, choosing the right marketing channels is essential to reaching your target audience effectively. Different marketing channels work better for different audiences, so it's important to determine which ones will be most effective for your specific customer base.

One of the most effective marketing channels for solar energy is digital marketing. This includes social media, email marketing, and online advertising. Digital marketing is a great way to reach a large audience quickly and efficiently, and can be highly targeted to specific demographics and interests.

However, traditional marketing techniques such as direct mail, radio and TV advertising, and billboards can still be effective for certain demographics, such as older or more rural populations. It's important to consider the specific characteristics of your target market and tailor your marketing efforts accordingly.

Another effective marketing channel for solar energy is events and community outreach. Hosting or participating in events such as solar fairs or educational seminars can help raise awareness about solar energy and attract potential customers. Community outreach, such as partnering with local organizations or sponsoring community events, can also be an effective way to build brand awareness and establish a positive reputation in the community.

Ultimately, the best marketing channels for your solar energy business will depend on the specific needs and characteristics of your target market. By understanding your customers and their behaviors, you can develop a marketing plan that effectively reaches them through the right channels and with the right messaging.

3. Digital Marketing For Solar Energy

In today's digital age, businesses that sell solar energy must embrace digital marketing to promote their products and services. Digital marketing provides businesses with powerful tools to reach potential customers and build brand awareness. Here are some effective digital marketing strategies to promote solar energy.

Firstly, having a website is essential. A website should be user-friendly, informative, and visually appealing. It should contain information about the benefits of solar energy, case studies, and customer testimonials. The website should also be optimized for search engines to improve visibility and drive traffic to the site.

Secondly, search engine optimization (SEO) is crucial. This process optimizes the website to rank higher in search engine results. By incorporating relevant keywords related to solar energy, businesses can improve their visibility in search engine results and drive more traffic to their website.

Thirdly, pay-per-click (PPC) advertising is a great way to reach potential customers. PPC advertising displays ads in search engine results and other websites, targeting specific keywords and demographics. It allows businesses to reach potential customers who are searching for information about solar energy.

Social media platforms such as Facebook, Twitter, and LinkedIn provide businesses with an excellent opportunity to reach potential customers and build brand awareness. By sharing informative and engaging content, businesses can establish themselves as thought leaders in the renewable energy sector.

Email marketing is a cost-effective way to reach potential customers and build brand awareness. Sending regular newsletters and promotional emails can keep potential customers informed about the benefits of solar energy and promote special offers.

Lastly, video marketing is a powerful way to showcase the benefits of solar energy and provide potential customers with a visual demonstration of how solar energy works. By creating informative and engaging videos, businesses can increase brand awareness and generate leads.

In conclusion, digital marketing is essential for businesses that sell solar energy. By having a user-friendly website, optimizing it for search engines, using PPC advertising, leveraging social media, using email marketing, and video marketing, businesses can reach potential customers and establish themselves as leaders in the renewable energy sector.

Building A Strong Online Presence

In the digital age, having a strong online presence is crucial for any business, including those in the solar energy industry. Building a strong online presence starts with creating a user-friendly website that is optimized for search engines. This can help potential customers find your business when they search for solar energy solutions online.

Another key aspect of digital marketing for solar energy is social media. Platforms like Facebook, Twitter, and Instagram provide a way to connect with potential customers and share information about your business and its offerings. By regularly posting informative content and engaging with followers, you can build a loyal following and increase brand awareness.

Email marketing is also a useful tool in the solar energy industry. By building a list of email subscribers and sending regular newsletters, you can keep customers informed about your business and its offerings, as well as share relevant news and updates about the industry.

Finally, online advertising can be an effective way to reach potential customers. Platforms like Google Ads and Facebook Ads allow you to target specific demographics and locations, ensuring that your ads are seen by the people who are most likely to be interested in your products or services.

Overall, digital marketing is an essential part of any successful solar energy marketing strategy. By building a strong online presence, businesses can reach a wider audience and connect with potential customers in new and innovative ways.

Utilizing Social Media For Reach And Engagement

When it comes to digital marketing for solar energy, social media is a powerful tool for reaching and engaging potential customers. With the right strategy, social media can help you build a strong online presence and establish your brand as a thought leader in the industry.

To get started, it's important to identify which social media platforms your target audience is using most. For example, if you're targeting a younger demographic, you may want to focus on platforms like Instagram and TikTok, while if you're targeting business owners, LinkedIn may be a more effective platform.

Once you've identified the best platforms for your target audience, it's important to create a content strategy that resonates with them. This may include sharing educational content about the benefits of solar energy, showcasing your projects and success stories, or providing tips for reducing energy costs.

One of the benefits of social media is the ability to engage with your audience in real-time. This means responding to comments and messages promptly and showing that you value their feedback. It also means being transparent and authentic in your communication, which can help to build trust and credibility with your audience.

To maximize the reach of your social media content, consider incorporating paid advertising into your strategy. This can help you reach a larger audience and target specific demographics based on factors like location, age, and interests.

Overall, social media is an important tool for digital marketing in the solar energy industry. By understanding your target audience and creating a content strategy that resonates with them, you can build a strong online presence and establish your brand as a leader in the industry.

Search Engine Optimization And Pay-Per-Click Advertising

Digital marketing can be an effective tool for promoting solar energy and reaching potential customers. Two key strategies for achieving this are search engine optimization (SEO) and pay-per-click (PPC) advertising.

SEO involves optimizing your website content and structure to improve your search engine rankings, making it more likely that potential customers will find your site when searching for relevant keywords. This can be achieved through techniques such as keyword research, on-page optimization, and link building.

PPC advertising, on the other hand, involves paying to have your ads displayed at the top of search engine results pages or on other relevant websites. This can be an effective way to quickly generate traffic and leads, especially when combined with targeted keywords and messaging.

To make the most of these strategies, it is important to understand your target audience and the keywords they are searching for. You can then tailor your content and messaging to these specific search terms, helping to improve your search engine rankings and generate more traffic to your site.

It is also important to continually monitor and optimize your SEO and PPC campaigns to ensure they are effectively reaching your target audience and generating a positive return on investment. By regularly analyzing your website traffic and conversion rates, you can make data-driven decisions and adjust your strategies as needed to improve your results over time.

In addition to SEO and PPC, social media can also be a valuable tool for promoting solar energy and engaging with potential customers. Platforms such as Facebook, Twitter, and LinkedIn can be used to share content, showcase customer success stories, and build relationships with followers. By building a strong online presence through these channels, you can establish yourself as a trusted resource for solar energy information and build a community of interested and engaged followers.

Facebook Ads And Facebook Groups

Social media is an integral part of any digital marketing strategy, and Facebook is no exception. With over 2.7 billion monthly active users, Facebook offers a massive potential audience for solar companies looking to promote their products and services.

One of the most effective ways to reach potential customers on Facebook is through targeted advertising. Facebook's powerful advertising platform allows you to create highly specific ads based on demographic information, interests, behaviors, and more. This means you can create ads that are tailored to your ideal customer, ensuring that your message is reaching the right people.

In addition to targeting, Facebook's ad platform offers a range of different ad formats to choose from. These include image ads, video ads, carousel ads, and more. Depending on your marketing goals and the needs of your target audience, you can choose the ad format that works best for your business.

One of the most appealing features of Facebook advertising is the ability to set a budget and track your ad performance in real-time. This means you can monitor the effectiveness of your ads and adjust your strategy accordingly to maximize your return on investment.

Another powerful tool for solar companies on Facebook is the use of Facebook groups. By creating or joining groups that are relevant to your industry or target audience, you can establish a community of engaged users who are interested in your products and services. This provides a valuable opportunity to share information, answer questions, and build relationships with potential customers.

Overall, Facebook advertising and groups can be an effective way for solar companies to reach their target audience and promote their products and services. By leveraging the power of Facebook's advertising platform and building an engaged community through groups, solar companies can increase their brand awareness, generate leads, and ultimately drive sales.

Want to learn to generate your own high quality leads using Facebook Ads? Have a look at The Solar King's Digital Marketing Course: https://thesolarking.net/solar-leads-course

Content Marketing And Email Marketing

When it comes to digital marketing for solar energy, content marketing and email marketing are two important strategies that can help businesses to connect with their target audience and drive conversions.

Content marketing involves creating and sharing valuable content that is designed to inform, educate, or entertain your target audience. This can include blog posts, videos, infographics, and other types of content that are relevant to the solar industry and address the needs and pain points of your potential customers.

By providing high-quality content that is useful and informative, businesses can establish themselves as thought leaders in the solar industry and build trust with their audience. This can help to attract potential customers and encourage them to take action, whether that means filling out a contact form or requesting a consultation.

Email marketing is another important strategy for solar businesses looking to connect with their audience and drive conversions. By building an email list of potential customers, businesses can send targeted messages that are personalized to the recipient's interests and needs.

This can include promotional offers, educational content, or updates on new products or services. By sending timely and relevant emails, businesses can keep their audience engaged and top-of-mind, increasing the likelihood that they will take action and become a customer.

Overall, content marketing and email marketing are two key strategies that can help solar businesses to reach and connect with their target audience. By providing valuable content and building relationships with potential customers, businesses can establish themselves as industry leaders and drive conversions over time.

4. Traditional Marketing Techniques For Solar Energy

While digital marketing has become increasingly popular, traditional marketing techniques can still be effective in promoting solar energy. There are various traditional marketing techniques that solar energy businesses can use to promote their services.

Print advertising is one such technique that can be used to reach potential customers through ads in newspapers and magazines, and the creation of flyers and brochures to distribute at local events.

Event marketing is another effective traditional marketing technique. By participating in trade shows, seminars, and community events, businesses can showcase their products and services, engage with potential customers, and build trust.

Direct mail marketing is another traditional marketing technique that businesses can use to reach potential customers who may not have internet access. Targeted mailings to specific neighborhoods or demographics can generate leads and promote services.

Radio and television advertising can also be a useful traditional marketing technique to reach a broad audience, while outdoor advertising through billboards and signs can capture the attention of potential customers who are driving or walking in the area.

Overall, while digital marketing has become increasingly popular, traditional marketing techniques such as print advertising, event marketing, direct mail marketing, radio and television advertising, and outdoor advertising can still be effective in promoting solar energy services and establishing businesses as experts in the renewable energy sector.

Networking And Building Partnerships

Traditional marketing techniques can be just as effective for promoting solar energy as digital marketing techniques. Networking and building partnerships with other businesses and organizations in the community can help to increase visibility and reach new potential customers.

One effective way to network and build partnerships is to attend community events and engage with other attendees. This can help to establish your business as a trusted member of the community and provide opportunities to meet potential customers and collaborators.

Another traditional marketing technique is to use print materials such as brochures and flyers to promote solar energy. These materials can be distributed at events, in local businesses, or through direct mail campaigns. They can be a great way to highlight the benefits of solar energy and provide potential customers with information on how to get started.

In addition, partnering with local organizations and businesses can help to expand your reach and tap into new customer bases. For example, partnering with a home improvement store can allow you to offer discounts or promotions to their customers, while partnering with a local environmental organization can help to promote the environmental benefits of solar energy.

Overall, networking and building partnerships through traditional marketing techniques can help to increase visibility, establish trust in the community, and reach new potential customers for your solar energy business.

Trade Shows And Events

Trade shows and events are traditional marketing techniques that can be very effective in promoting solar energy products and services. These events provide a platform for businesses to showcase their products and services to a wider audience and to meet potential customers face to face.

Trade shows and events related to solar energy are attended by a range of people, including industry experts, policymakers, investors, and consumers. These events provide an opportunity to showcase the latest solar technologies, discuss policy issues, and connect with potential customers.

Participating in trade shows and events can be an excellent way to build brand awareness, generate leads, and establish new partnerships. For businesses in the solar industry, these events offer a unique opportunity to reach a large and diverse audience and to promote their products and services in a more personal and interactive way.

When preparing for a trade show or event, it is important to have a clear message and a well-designed booth that highlights the key features and benefits of your product or service. Engaging attendees with interactive displays, product demonstrations, and presentations can also help to create a memorable experience and generate interest in your business.

By participating in trade shows and events, solar energy businesses can stay up to date on the latest trends and developments in the industry, connect with potential customers, and promote their products and services in a more personal and engaging way.

Door Knocking

Door knocking is a traditional marketing technique that involves going door-to-door to engage with potential customers. This approach can be effective for solar energy sales, especially when targeting homeowners in a specific area. Door knocking allows you to engage with potential customers face-to-face, answer their questions, and address any concerns they may have about solar energy.

When door knocking, it's important to be respectful and courteous to the homeowner. Start by introducing yourself and explaining the purpose of your visit. Let them know that you are there to talk about solar energy and the benefits it can provide for their home. Be prepared to answer questions and provide information about the benefits of solar energy, such as lower energy bills and reduced carbon emissions.

It's important to keep in mind that not everyone will be interested in solar energy, and that's okay. Some people may already have solar panels or may not be in a position to invest in solar energy at the moment. However, by engaging with potential customers through door knocking, you can increase awareness of solar energy and its benefits, which may lead to future sales or referrals.

Door knocking can be a time-consuming and labor-intensive approach to marketing solar energy, but it can be an effective way to engage with potential customers and build relationships in the community. With the right approach and messaging, door knocking can be a valuable tool in your solar energy marketing toolkit.

If you would like to see a free training video on how to knock doors in a much more efficient way have a look at this video:

https://thesolarking.net/door-knocking

Direct Mail And Print Advertising

Direct mail and print advertising are traditional marketing techniques that can be effective for promoting solar energy. Direct mail involves sending marketing materials directly to potential customers' mailboxes, while print advertising involves placing ads in newspapers, magazines, and other print publications.

Direct mail campaigns can be targeted to specific demographics, making them a useful tool for reaching potential customers who may be interested in solar energy. Direct mail materials can include flyers, brochures, and postcards that highlight the benefits of solar energy and promote the products and services offered by the solar company.

Print advertising can also be an effective way to reach a broad audience. Ads can be placed in local newspapers, magazines, and other print publications that are likely to be read by the target demographic. Print ads can be used to build brand recognition, promote specific products or services, and drive traffic to the company's website.

When using direct mail and print advertising, it is important to craft compelling messaging and use eye-catching visuals to grab the reader's attention. It is also important to track the success of these campaigns and adjust the messaging and visuals as needed to improve their effectiveness.

While digital marketing has become increasingly popular, traditional marketing techniques like direct mail and print advertising can still be an effective way to promote solar energy and reach potential customers. By using a mix of both traditional and digital marketing techniques, solar companies can maximize their reach and connect with a wider audience.

Radio And Television Advertising

Radio and television advertising are traditional marketing techniques that can be effective in promoting solar energy. While digital marketing has become increasingly popular, radio and television still offer a wide reach and can be particularly useful for targeting local audiences.

One advantage of radio advertising is its ability to target specific demographics. Radio stations often have a specific listener demographic, which can be a useful tool for solar companies to reach their target audience. Additionally, radio advertising can be relatively inexpensive and can be a cost-effective option for smaller solar companies.

Television advertising, on the other hand, can be a more expensive option, but offers the benefit of reaching a larger audience. Television advertising can be particularly effective for solar companies looking to build brand recognition and awareness. Additionally, television advertising can be used to promote specific campaigns, promotions or events, such as a community solar program or a solar panel installation promotion.

Both radio and television advertising can be effective when paired with other traditional and digital marketing techniques. For example, a television or radio ad can be used to promote a solar company's website or social media accounts.

Overall, while traditional marketing techniques like radio and television advertising are often considered less effective than digital marketing, they can still be a valuable tool for solar companies looking to reach local audiences or build brand recognition. By understanding the advantages and limitations of different marketing techniques, solar companies can develop a well-rounded marketing strategy that effectively promotes their products and services.

5. Measuring The Success Of Your Solar Marketing Efforts

To measure the success of your solar marketing efforts, it's important to analyze several key factors. Firstly, monitoring website traffic can provide insight into how many people are engaging with your content. This information can help you identify areas that need improvement on your website and adjust your marketing strategies accordingly.

Conversion rates are another important factor to measure. This refers to the percentage of visitors who take a desired action on your website, such as filling out a contact form or making a purchase. Tracking conversion rates can help you identify areas where potential customers may be getting stuck in the sales process, and make changes to improve conversion rates.

Lead generation is also a crucial metric to track. This involves analyzing the number of leads generated through your marketing efforts to determine the effectiveness of your campaigns. By identifying which marketing channels generate the most leads, you can focus your efforts on the channels that are most effective.

Engagement on social media is another important factor to measure. This includes tracking likes, shares, and comments on social media platforms, which can provide valuable insights into the success of your social media marketing efforts.

Customer feedback is also valuable in determining the effectiveness of your marketing strategies. Gathering feedback from customers can help you identify areas where you may need to make changes or improvements to better meet the needs of your customers.

Finally, it's important to measure the return on investment (ROI) of your marketing efforts. This involves analyzing the cost of your marketing campaigns compared to the revenue generated. By determining which campaigns are most effective, you can adjust your strategies to improve overall marketing success.

Overall, measuring the success of your solar marketing efforts involves analyzing website traffic, conversion rates, lead generation, social media engagement, customer feedback, and ROI. By monitoring these factors, you can identify areas of strength and weakness in your marketing campaigns and make data-driven decisions to improve your overall marketing success.

Setting Metrics For Measuring Success

Measuring the success of your solar marketing efforts is important to understand the effectiveness of your marketing strategies and make any necessary adjustments. But how do you measure success in solar marketing?

First, it's essential to define your goals and objectives. What do you want to achieve through your marketing efforts? Is it to increase website traffic, generate leads, or boost sales? Once you have established your objectives, you can then set metrics to measure progress towards those goals.

One of the most common metrics used in digital marketing is website analytics. Tools like Google Analytics can provide data on website traffic, page views, bounce rates, and other user behavior metrics. By analyzing this data, you can determine which pages on your website are most engaging to users, which campaigns are driving the most traffic, and how long visitors are spending on your website.

Another key metric to consider is lead generation. By tracking the number of leads generated through various marketing channels, you can evaluate which strategies are most effective and allocate resources accordingly. You can use lead scoring to determine the quality of leads generated and prioritize those that are most likely to convert into paying customers.

Measuring the success of your social media marketing efforts can be more challenging, but it is still possible. You can use tools like Facebook Insights and Twitter Analytics to track engagement, reach, and follower growth. By analyzing these metrics, you can identify which social media platforms are most effective for reaching your target audience and adjust your strategy accordingly.

Ultimately, the success of your solar marketing efforts will depend on your ability to continually adapt and improve your strategies. By regularly monitoring and analyzing data, you can identify areas for improvement and optimize your marketing efforts for maximum impact.

Analyzing Data And Adjusting Your Strategy As Needed

Measuring the success of your solar marketing efforts is critical to ensure that you are reaching your target audience and achieving your marketing goals. By analyzing data and adjusting your strategy as needed, you can optimize your marketing efforts and maximize your return on investment. Here are some key steps to help you measure the success of your solar marketing efforts:

Set metrics for measuring success: Before you start your marketing campaign, it is important to define your metrics for success. This may include metrics such as website traffic, leads generated, conversions, and social media engagement. By setting clear metrics, you can easily track your progress and determine if your marketing efforts are achieving your goals.

Use analytics tools to track your metrics: There are many tools available to help you track your marketing metrics, such as Google Analytics, social media analytics tools, and marketing automation software. These tools can help you gather data and insights about your target audience and their behavior, allowing you to better understand how they are engaging with your marketing campaigns.

Analyze the data: Once you have collected your marketing data, it is important to analyze it to gain insights into your target audience's behavior and preferences. This can help you identify areas for improvement in your marketing strategy and optimize your future marketing efforts.

Adjust your strategy: Based on your data analysis, make adjustments to your marketing strategy to better align with the needs and preferences of your target audience.

This may include revising your messaging, targeting specific demographics, or adjusting your marketing channels.

Continuously track your metrics: Once you have made adjustments to your marketing strategy, continue to track your metrics to determine if your changes have resulted in improved performance. By continuously tracking and analyzing your marketing data, you can optimize your marketing efforts over time and ensure that you are achieving your marketing goals.

Overall, measuring the success of your solar marketing efforts is critical to ensuring that you are effectively reaching your target audience and achieving your marketing goals. By setting clear metrics, using analytics tools, analyzing your data, adjusting your strategy, and continuously tracking your metrics, you can optimize your marketing efforts and maximize your return on investment.

Continuous Improvement And Testing

In the solar energy industry, measuring the success of your marketing efforts is crucial for your business growth. The key to improving your marketing strategy is to continuously track and analyze the data collected, and to make necessary changes to your approach as needed. This is where the process of continuous improvement and testing comes into play.

One of the most important things to keep in mind when measuring the success of your solar marketing efforts is to have a clear understanding of your marketing goals and objectives. By having well-defined goals, you can establish the metrics needed to measure the success of your marketing campaigns. Some common metrics include website traffic, leads generated, customer acquisition costs, and conversion rates.

To effectively analyze your data, it's important to use the right tools and resources. Many digital marketing tools offer features such as analytics, tracking, and reporting that can help you measure the success of your campaigns. With these tools, you can identify trends, strengths, weaknesses, and areas for improvement. You can then make data-driven decisions to adjust your marketing strategies and tactics accordingly.

Another important aspect of measuring the success of your solar marketing efforts is continuous improvement and testing. This involves regularly testing different marketing channels, tactics, and messaging to find the most effective combination for your target audience. By testing and tweaking your campaigns, you can improve your overall marketing strategy and achieve better results.

It's important to remember that measuring the success of your solar marketing efforts is an ongoing process. By constantly tracking and analyzing data, adjusting your strategy as needed, and testing new approaches, you can continue to improve your marketing efforts and grow your business in the solar energy industry.

The Solar King's Advice to Solar Marketing Professionals

When working on a solar marketing campaign, whether it is digital or any other medium, it is important to remember what you like is not necessarily what will convert to leads and ultimately closed deals.

Testing is one of the most important aspects of marketing. Images, text, colors, etc.. are all variables that may be tested in different formats and mediums. Once you develop a winning campaign it is important to continue testing and developing new winners because you never know when any given campaign will stop working.

My company grew to be one of the top solar marketing agencies in the United States because I was constantly innovating and trying new things while my competition often stopped trying to get better and relied on marketing campaigns that sometimes continued to work for years.

I encourage you to never stop testing and trying new ideas. You will be shocked with the endless marketing campaigns you can come up with that will convert to leads and closed deals.

The Solar King's Additional Resources For Solar Marketers And Sales Organizations

- The Solar King's Free Solar Marketing Resources
  https://thesolarking.net/free-solar-resources
- The Solar King's Digital Marketing Course
  https://thesolarking.net/solar-leads-course
- The Solar King's Done For Your Marketing Package:
  https://thesolarking.net/offer
- The Solar King's Coaching Programs
  https://thesolarking.net/coaching
- The Solar King's Live Training Events
  https://thesolarking.net/events

Selling Solar Energy

1. The Benefits Of Selling Solar Energy

Selling solar energy provides a range of benefits for businesses that specialize in this area. The benefits of selling solar energy include not only financial rewards, but also the satisfaction of contributing to a more sustainable future and helping customers achieve energy independence. Here are some of the key benefits of selling solar energy:

First and foremost, selling solar energy can be a profitable business opportunity. As the demand for renewable energy solutions grows, there is a growing market for solar energy products and services. By positioning themselves as leaders in the renewable energy sector, businesses can attract new customers and establish themselves as trusted experts in the field.

Selling solar energy can also provide opportunities for innovation and creativity. As solar technology continues to improve, there are many opportunities for businesses to develop new products and services that meet the needs of their customers. By staying up-to-date on the latest developments in solar technology, businesses can position themselves as innovative and forward-thinking.

In addition to financial rewards and opportunities for innovation, selling solar energy also provides the satisfaction of contributing to a more sustainable future. By helping households and businesses adopt renewable energy solutions, businesses that sell solar energy are playing an important role in reducing our collective carbon footprint and mitigating climate change.

Finally, selling solar energy also provides the opportunity to help customers achieve energy independence. By generating their own electricity, customers can reduce their dependence on the traditional power grid and become more self-sufficient. This can help to ensure a reliable source of electricity, particularly during power outages and other disruptions.

In order to reap the benefits of selling solar energy, businesses must be knowledgeable about solar technology and be able to provide high-quality products and services to their customers. By educating themselves and their customers about the benefits of solar energy and providing exceptional customer service, businesses can build a loyal customer base and establish themselves as trusted experts in the field.

Overall, selling solar energy provides a range of benefits for businesses, including financial rewards, opportunities for innovation, and the satisfaction of contributing to a more sustainable future. By emphasizing the economic, environmental, and energy independence benefits of solar energy, businesses that sell solar energy can attract new customers and help to create a more sustainable and equitable future.

High Profit Potential

Selling solar energy can be a profitable venture for businesses and individuals alike. One of the key benefits of selling solar energy is the high profit potential. As more people become interested in renewable energy, the demand for solar panels is growing. This demand presents a unique opportunity for businesses and individuals to capitalize on the growing market.

Selling solar energy can also be an effective way to differentiate oneself from the competition. By offering a unique and in-demand product, businesses and individuals can set themselves apart from others in their industry. This can lead to increased brand recognition and customer loyalty, ultimately resulting in more sales and long-term success.

Another benefit of selling solar energy is the positive impact it can have on the environment. By promoting and selling renewable energy, businesses and individuals are contributing to a more sustainable future. This can lead to a positive brand image and an increased reputation among environmentally conscious consumers.

Selling solar energy can also bring about a sense of fulfillment for those who are passionate about renewable energy and making a positive impact on the world. Knowing that one's work is contributing to a more sustainable future can be incredibly rewarding, and can lead to a greater sense of purpose and satisfaction in one's work.

Overall, the benefits of selling solar energy include high profit potential, differentiation from competitors, positive impact on the environment, and a sense of fulfillment. By understanding these benefits and taking advantage of the growing market for renewable energy, businesses and individuals can succeed in selling solar energy and contribute to a more sustainable future.

Job Creation And Economic Growth

The benefits of selling solar energy extend beyond just the potential for high profits. The solar industry is a rapidly growing industry, and with growth comes job creation and economic growth. As more households and businesses adopt solar energy, the demand for solar products and services increases, leading to the creation of more jobs in manufacturing, installation, maintenance, and research.

In addition to job creation, the solar industry can contribute to overall economic growth. The installation of solar panels and related equipment can stimulate local economies, as materials are purchased from local suppliers, and workers spend their paychecks in the community. Moreover, solar energy can lead to increased property values and lower energy costs, which can positively impact the financial well-being of homeowners and businesses.

Furthermore, the solar industry has the potential to attract and retain a talented workforce. As the industry continues to grow, it requires a diverse set of skilled professionals, from engineers and project managers to sales representatives and customer service specialists. The industry's focus on sustainability and clean energy can also make it an attractive option for individuals who are passionate about environmental issues and making a positive impact.

Overall, the benefits of selling solar energy extend far beyond just financial gains. The industry's potential for job creation, economic growth, and attracting a talented workforce make it an appealing option for those looking to make a positive impact on their community and the environment.

Positive Impact On The Environment

Selling solar energy offers a range of benefits, including the potential for significant positive impact on the environment. By helping households and businesses to transition away from fossil fuels, solar energy can reduce carbon emissions and help combat climate change. This can have far-reaching benefits, including cleaner air and water, and a healthier environment for both people and wildlife.

Another major benefit of selling solar energy is the positive impact on the economy. The solar industry creates jobs and drives economic growth, providing opportunities for people across a range of fields, from manufacturing and installation to research and development. This can help to build strong, vibrant communities and support a healthy economy.

In addition to economic and environmental benefits, selling solar energy can also provide a range of other advantages, such as increased energy independence, improved public health, and reduced reliance on traditional power sources. By helping households and businesses to generate their own electricity, solar energy can provide greater control over energy costs and reduce exposure to price fluctuations in the energy market.

Overall, selling solar energy can be a rewarding and lucrative business opportunity that offers numerous benefits for both the seller and the customer. By helping to promote a cleaner, more sustainable future and supporting economic growth and job creation, those who sell solar energy can make a positive impact on both the environment and their communities.

Opportunities For Innovation And Growth In The Industry

The solar energy industry offers a vast array of opportunities for innovation and growth, making it an exciting field to be in. The development of new technologies and the need for continuous improvement mean that there is always room for growth and advancement. By being involved in the solar energy industry, individuals can be a part of shaping the future of sustainable energy.

Selling solar energy also offers opportunities for business growth and expansion. As more households and businesses adopt solar energy, the demand for products and services related to solar energy is also increasing.

This means that there is a growing market for solar energy products and services, creating new opportunities for businesses to expand and grow.

Furthermore, the solar energy industry is a source of job creation and economic growth. As the industry grows, it creates new jobs in areas such as manufacturing, installation, maintenance, and research. This means that by selling solar energy, individuals and businesses are not only contributing to a more sustainable energy future, but also to the growth of the economy and job creation in their community.

In addition, selling solar energy has a positive impact on the environment. By promoting the use of solar energy, individuals and businesses can help reduce carbon emissions and promote a more sustainable future. This not only benefits the environment but also creates a sense of social responsibility and environmental awareness.

Benefits Of Adding Solar Sales To Existing Home Improvement Companies

---

Adding solar sales to home improvement companies, such as roofing and HVAC businesses, can provide numerous benefits. By offering solar options, these companies can expand their services and provide a more comprehensive approach to meeting their customers' energy needs. This can lead to increased customer satisfaction and loyalty, as well as additional revenue streams for the business.

Moreover, the overlap between the industries of roofing and solar is significant, as many solar panel installations are integrated into the roof of a home or business. This means that roofing companies already have much of the infrastructure and expertise needed to install solar panels. By adding solar sales to their business, these companies can leverage their existing skills and experience to expand into a growing and profitable industry.

Home improvement companies can also help to promote the adoption of solar energy and contribute to a more sustainable future. As trusted providers of energy-related services, they are well-positioned to educate their customers about the benefits of solar and encourage them to make the switch. This can not only benefit the environment but also help to save customers money on their energy bills. Overall, adding solar sales to their business can be a win-win for home improvement companies and their customers alike.

Overall, selling solar energy offers numerous benefits, including high profit potential, opportunities for innovation and growth, job creation and economic growth, and a positive impact on the environment. By being a part of the solar energy industry, individuals and businesses can contribute to a more sustainable future while also growing and expanding their own businesses.

Learn how roofing companies can add a solar division to their business within seven days for $0 by working with The Solar King:

https://thesolarking.net/roofing

2. The Skills And Qualifications Needed For Selling Solar Energy

Selling solar energy requires a range of skills and qualifications in order to be successful. In addition to a solid understanding of solar technology and the benefits of renewable energy, salespeople in the solar industry must also possess strong communication skills, a customer-focused mindset, and the ability to develop and maintain relationships with clients.

One of the most important skills for selling solar energy is a deep understanding of the technology and the benefits of renewable energy. Salespeople must be able to explain the benefits of solar energy in a clear and compelling way, and must be able to answer any questions that potential customers may have. This requires a strong technical knowledge of solar technology, including how solar panels work, how they are installed, and how they generate electricity.

In addition to technical knowledge, salespeople must also possess strong communication skills. They must be able to communicate the benefits of solar energy in a clear and compelling way, and must be able to adapt their communication style to meet the needs of different clients. This requires strong listening skills, the ability to ask questions, and the ability to explain complex concepts in a simple and understandable way.

Another important skill for selling solar energy is a customer-focused mindset. Salespeople must be able to understand the needs and priorities of their clients, and must be able to develop and present solar energy solutions that meet those needs. This requires a deep understanding of the customer's business or household, as well as the ability to develop and maintain strong relationships with clients.

Finally, salespeople in the solar industry must be able to work effectively as part of a team. They must be able to collaborate with technical experts and other salespeople in order to develop and present the best possible solar energy solutions for their clients. This requires strong teamwork skills, the ability to work well under pressure, and the ability to adapt to changing circumstances.

Overall, selling solar energy requires a range of skills and qualifications, including technical knowledge, communication skills, a customer-focused mindset, and the ability to work effectively as part of a team. By developing these skills and qualifications, salespeople can be successful in the solar industry and help to create a more sustainable and equitable future.

Technical Knowledge Of Solar Energy Systems

Selling solar energy requires a unique set of skills and qualifications to effectively educate customers and close deals. One important qualification is technical knowledge of solar energy systems. Solar sales professionals need to be well-versed in the technical aspects of solar energy, including the different types of solar panels, inverters, and battery storage systems. They should be able to explain to customers how these systems work and answer any questions they may have.

In addition to technical knowledge, effective sales professionals in the solar industry need to have strong communication and interpersonal skills. They should be able to explain complex technical concepts in a way that is easy for the customer to understand. They should also be able to listen to the customer's needs and concerns and provide solutions that meet their specific needs.

A background in renewable energy or engineering can also be beneficial for a career in solar sales. However, a passion for renewable energy and the ability to learn quickly can also go a long way in this field.

Sales professionals in the solar industry should also be knowledgeable about industry regulations and policies, as well as the financial incentives available to customers. They should be able to help customers navigate the incentives and financing options available to make the purchase of solar energy systems more affordable.

Finally, sales professionals in the solar industry should be committed to providing excellent customer service and follow-up. They should be available to answer any questions or concerns that may arise after the installation of a solar energy system and be able to provide ongoing support to ensure the system is running efficiently.

Overall, the skills and qualifications needed for selling solar energy require a combination of technical knowledge, communication and interpersonal skills, industry knowledge, and a commitment to excellent customer service. With these skills, sales professionals can effectively educate and close deals with customers, promoting the growth of the solar industry and a more sustainable future.

Sales And Marketing Skills

Any salesperson needs to be able to identify customer needs and pain points, offer customized solutions, and effectively address any objections or concerns that may arise. Additionally, strong communication and negotiation skills are critical for closing the sale.

Marketing skills are also important for selling solar energy. A solar salesperson should be able to create and execute effective marketing campaigns that reach their target audience. They need to be able to identify the best marketing channels to use for their target market, develop messaging that resonates with their audience, and create compelling content that drives engagement.

In addition to sales and marketing skills, technical knowledge of solar energy systems is also essential. A salesperson needs to be able to explain the technical aspects of solar energy in a way that is easy for customers to understand. They need to be knowledgeable about different solar panels and inverters, as well as energy storage solutions and other related technologies.

Overall, selling solar energy requires a combination of technical knowledge, sales skills, and marketing expertise. The ability to effectively communicate the benefits of solar energy and address customer concerns is essential for success in the industry. With the right skills and qualifications, a solar salesperson can help customers make the switch to clean, renewable energy while building a successful career in the solar industry.

Understanding Of The Local Regulatory Environment

Selling solar energy also requires an understanding of the local regulatory environment. Technical knowledge is important because potential customers will look to you for advice on what solar energy system would work best for their property, and they will have questions about the installation process, maintenance, and performance. You'll need to be able to confidently answer those questions and provide accurate information.

Sales and marketing skills are also essential. You'll need to be able to communicate the benefits of solar energy to potential customers in a way that is persuasive and convincing. This involves listening to their needs and concerns and tailoring your pitch to their specific situation. You'll also need to be able to build relationships with potential customers and follow up with them to address any questions or concerns they may have.

Finally, an understanding of the local regulatory environment is important for navigating the legal requirements and incentives that may be available. This includes knowledge of state and local regulations around solar energy systems, as well as any tax credits or other incentives that may be available. By staying up-to-date on the regulatory environment, you can offer your customers the best advice and guidance, and help them take advantage of any available incentives.

Overall, selling solar energy requires a diverse set of skills and qualifications. With the right combination of technical knowledge, sales and marketing skills, and an understanding of the local regulatory environment, you can build a successful career in this growing industry.

Networking And Business Development Skills

Selling solar energy is also about sales and marketing skills. You need to be able to identify potential customers, understand their needs and pain points, and craft your messaging and pitch to appeal to their interests. You need to be able to explain the benefits of solar energy in a clear and persuasive way, and be able to handle objections and concerns.

Understanding the local regulatory environment is also important, as different regions may have different rules and regulations regarding the installation of solar energy systems. You need to be familiar with the permits and approvals that are required, and be able to guide your customers through the process.

Finally, networking and business development skills are also important. You need to be able to build relationships with potential customers, industry partners, and other stakeholders in the solar energy industry. This will help you to identify new opportunities for growth and build a strong network of contacts that can support your business.

In summary, the skills and qualifications needed for selling solar energy are technical knowledge of solar energy systems, sales and marketing skills, an understanding of the local regulatory environment, and networking and business development skills. By developing these skills, you can build a successful career in the solar energy industry and help to promote the growth and adoption of this important technology.

3. Overcoming Challenges In Selling Solar Energy

Selling solar energy can be a challenging endeavor, as it requires navigating a complex and rapidly changing landscape. From regulatory hurdles to technical issues to customer skepticism, businesses that sell solar energy must be prepared to overcome a range of challenges in order to be successful. Here are some of the key challenges faced by businesses that sell solar energy, and strategies for overcoming them:

One of the biggest challenges in selling solar energy is navigating the complex regulatory environment. Regulations and incentives for solar energy can vary widely by state and by country, and can change frequently. Businesses must be able to stay up-to-date on regulatory changes and must be able to navigate complex permit processes in order to install solar energy systems. One strategy for overcoming regulatory challenges is to partner with local experts who have experience navigating the regulatory landscape, and to develop relationships with local officials and policymakers.

Another challenge in selling solar energy is customer skepticism. Many customers are skeptical of the benefits of solar energy, and may be reluctant to make the switch due to concerns about cost, reliability, and technical issues. One strategy for overcoming customer skepticism is to provide clear and transparent information about the benefits of solar energy, and to offer educational materials that explain how solar energy systems work and how they can benefit the customer. Providing customer testimonials and case studies can also be an effective way to demonstrate the benefits of solar energy and build trust with potential customers.

Technical issues can also be a challenge in selling solar energy. Solar energy systems can be complex and require specialized knowledge and expertise to install and maintain. Businesses must be able to provide high-quality installation and maintenance services in order to ensure that solar energy systems operate effectively and efficiently. One strategy for overcoming technical challenges is to invest in training and development for installation and maintenance staff, and to partner with local technical experts who can provide support and guidance as needed.

Finally, financing can be a significant challenge for businesses that sell solar energy. Solar energy systems can be expensive, and many customers may not have the resources to pay for them upfront. Businesses must be able to offer financing options that make solar energy systems accessible to a wide range of customers. One strategy for overcoming financing challenges is to partner with local banks and financial institutions to offer low-interest loans and other financing options, and to develop creative financing solutions that meet the needs and priorities of different customer segments.

Overall, selling solar energy can be a challenging but rewarding endeavor. By staying up-to-date on regulatory changes, addressing customer skepticism, investing in training and development, and offering creative financing solutions, businesses can overcome the challenges of selling solar energy and help to create a more sustainable and equitable future.

Competition From Established Players In The Industry

As the demand for solar energy continues to grow, so does the competition among industry players. This can pose a challenge for those looking to enter the market and establish themselves as a viable option for customers. However, with the right strategies, it's possible to overcome this challenge and succeed in selling solar energy.

One key approach is to differentiate your offerings and highlight your unique value proposition. This may involve developing innovative products or services, leveraging advanced technology, or focusing on specific customer needs that are not being met by competitors. By standing out in the market and offering something unique, you can attract customers who are looking for more than just a run-of-the-mill solar provider.

Another approach is to build a strong brand and reputation through effective marketing and public relations efforts. By building trust and credibility with potential customers, you can establish yourself as a reliable and reputable solar provider that they can count on for their energy needs. This can involve developing a strong online presence, participating in community events, and leveraging customer testimonials and referrals to showcase your value.

Another challenge in selling solar energy is the high cost of installation and equipment. To overcome this challenge, you can offer financing options and incentives that make solar energy more accessible and affordable for a wider range of customers. This may involve partnering with financial institutions or government programs to offer low-interest loans or rebates that help offset the upfront costs of installation.

Overall, the key to overcoming competition and other challenges in selling solar energy is to focus on your unique value proposition, build a strong brand and reputation, and offer financing and other incentives that make solar energy more accessible and affordable for customers. By staying innovative and adaptable, you can succeed in this growing industry and help promote a more sustainable energy future.

Navigating The Regulatory Environment

The need to navigate complex and ever-changing regulations continues to grow as solar becomes more popular. From federal and state tax incentives to local zoning laws and building codes, the regulatory environment can be a significant barrier to selling solar energy.

One way to overcome this challenge is to stay informed and up-to-date on regulatory changes that may impact your business. This can involve regularly monitoring industry publications and attending relevant conferences or events. It may also involve building relationships with local officials and lawmakers to better understand the regulatory landscape and advocate for favorable policies.

Another approach is to partner with experienced professionals who can provide guidance and support in navigating the regulatory environment.

This may include lawyers who specialize in energy law or consultants who are well-versed in local regulations and permitting processes.

Ultimately, overcoming the regulatory challenges associated with selling solar energy requires a combination of knowledge, persistence, and creativity. By staying informed and building strong partnerships, solar energy businesses can successfully navigate the regulatory landscape and achieve their goals.

Addressing Customer Concerns And Objections

Selling solar energy can present some challenges, especially when it comes to addressing customer concerns and objections. One of the most common concerns customers may have is the upfront cost of installing solar panels. Many customers may be hesitant to make a large investment in solar energy without a clear understanding of the long-term benefits and savings.

To address this concern, it's important to provide customers with a clear understanding of the financial benefits of solar energy, such as reduced energy bills and increased property values. It's also important to provide customers with financing options and to explain how the long-term savings can offset the initial investment.

Another common objection that customers may have is the reliability and performance of solar energy systems. It's important to address these concerns by providing customers with information about the quality and durability of solar panels, as well as information about the performance of solar energy systems in their specific area.

Other concerns may include maintenance and repair costs, as well as the aesthetics of solar panels. By addressing these concerns and objections in a clear and honest manner, and providing customers with the information they need to make an informed decision, it's possible to overcome these challenges and build a successful business selling solar energy.

Staying Ahead Of Technological Advancements In The Industry

As the solar industry continues to evolve and improve, it's essential for those selling solar energy to stay ahead of technological advancements. This can be a significant challenge, as technology is always changing, and it can be difficult to keep up with the latest innovations.

One way to overcome this challenge is to stay informed and educated about new developments in the industry. This can include attending conferences and trade shows, participating in training programs and webinars, and reading industry publications.

Another way to stay ahead of technological advancements is to build partnerships and relationships with manufacturers and suppliers who are at the forefront of solar technology. This can provide valuable insights into new products and services that can help you better meet the needs of your customers.

It's also essential to invest in research and development, testing and trialing new technologies to determine what works best in different scenarios. By being proactive and continuously experimenting, you can stay at the forefront of the industry and ensure that you are providing the most up-to-date solutions to your customers.

Ultimately, staying ahead of technological advancements is crucial for those selling solar energy. By keeping up-to-date with new developments, building relationships with industry leaders, and investing in research and development, you can better serve your customers and stand out in a highly competitive market.

4. Closing The Sale In Solar Energy Marketing

Closing the sale is the ultimate goal of any solar energy sales and marketing campaign. It is the process of convincing potential customers to invest in solar energy products and services. Here are some effective strategies for closing the sale:

Building trust is crucial in the sales process. Providing transparent information about the products and services offered, answering questions honestly, and providing excellent customer service can help establish trust with potential customers.

Addressing concerns is also an important part of the sales process. Potential customers may have concerns about the cost, installation process, and maintenance of solar energy systems. Providing clear explanations and addressing these concerns can help to alleviate fears and convince customers that solar energy is a wise investment.

Offering incentives such as discounts, rebates, or financing options can make solar energy systems more affordable and attractive to potential customers.

Providing exceptional service throughout the sales process, from initial contact to post-installation follow-up, can help build trust and loyalty with customers.

Following up with potential customers who have expressed interest in solar energy but have not yet made a purchase can be a highly effective strategy. A simple phone call or email can remind them of the benefits of solar energy and encourage them to move forward with a purchase.

Providing clear and accurate information about the benefits of solar energy, installation process, and warranties can help to convince potential customers that investing in solar energy is a wise choice.

Including a clear call-to-action, such as a deadline for an incentive or a special promotion, can create a sense of urgency and motivate potential customers to make a purchase.

Overall, closing the sale involves building trust, addressing concerns, offering incentives, providing exceptional service, following up, providing clear information, and including a clear call-to-action. By using these strategies, solar energy businesses can effectively convince potential customers to invest in solar energy systems and create a more sustainable future.

Want to get my best resources to help you close more deals for free? Get them at this link:

https://thesolarking.net/free-solar-resources

Overcoming Customer Objections

Closing the sale is a crucial part of solar energy marketing. To close a sale, you need to understand the customer's objections and address them effectively. Common objections to purchasing solar energy include cost, reliability, and aesthetics.

To overcome cost objections, it's important to highlight the long-term financial benefits of installing solar energy. While the initial investment may be higher, the savings on energy bills and potential income from excess energy sold back to the grid can offset the cost over time. Offering financing options and incentives can also make solar energy more accessible and affordable for customers.

Reliability is another common concern, but it's important to emphasize the durability and reliability of solar energy systems. With proper installation and maintenance, solar panels can last for decades and continue to provide energy even in adverse weather conditions.

Aesthetics can also be a concern, especially for homeowners who want their property to look good. Fortunately, advances in solar panel design have made them more aesthetically pleasing, with options for sleek and streamlined panels that blend in with the roof.

To close the sale, it's important to address these objections head-on and provide solutions that address the customer's specific concerns. This requires active listening and a deep understanding of the customer's needs and priorities.

It's also important to provide clear and concise information about the installation process, timeline, and ongoing maintenance requirements. This helps build trust and confidence in the customer, and ensures a smooth and successful installation.

In summary, closing the sale in solar energy marketing requires a deep understanding of the customer's objections and priorities, and a commitment to addressing them effectively. By providing clear and concise information and offering solutions that meet the customer's needs, you can build trust and confidence, and ultimately close the sale.

Building Trust And Credibility

When it comes to closing the sale in solar energy marketing, building trust and credibility with potential customers is crucial. Solar energy is a significant investment, and customers need to feel confident in the quality and reliability of the products and services being offered.

One way to build trust and credibility is by providing testimonials and case studies from satisfied customers. This demonstrates to potential customers that the product or service has been successful for others, and can be successful for them as well. Additionally, featuring customer success stories on the company website and social media channels can help to establish the brand as a trustworthy and credible source in the industry.

Another key factor in building trust is providing transparent information about the products and services being offered. Customers want to understand exactly what they are paying for and what benefits they can expect to receive. Clear and honest communication about the costs, savings, and environmental benefits of solar energy can help to build trust with potential customers.

It's also important to address any concerns or objections that potential customers may have. Common objections include concerns about the cost of installation and maintenance, doubts about the effectiveness of solar energy in certain weather conditions, and uncertainty about the long-term reliability of the equipment.

By addressing these concerns directly and providing clear information about the benefits and potential cost savings of solar energy, companies can help to overcome these objections and close the sale.

In summary, building trust and credibility, providing transparent information, and addressing customer concerns are key elements of successfully closing the sale in solar energy marketing. By focusing on these factors, companies can establish themselves as trustworthy and reliable sources in the industry and build long-lasting relationships with satisfied customers.

Offering Incentives And Financing Options

To close a sale, a solar energy salesperson must convince a potential customer to make a purchase. One effective way to do this is by offering incentives and financing options. Incentives can include tax credits, rebates, or other financial benefits that reduce the cost of solar panel installation. Financing options such as leases, loans, or power purchase agreements can help make solar energy more affordable for customers who may not have the upfront funds to purchase a solar system outright.

Another key element in closing the sale is building trust and credibility with the customer. A solar energy salesperson can achieve this by providing detailed and accurate information about the benefits of solar energy and the installation process. This can include information about energy savings, the environmental benefits of solar, and the long-term financial benefits of solar panel installation. Providing references from satisfied customers can also help build trust and credibility with potential buyers.

Overcoming customer objections is another important aspect of closing a sale. Common objections to solar panel installation may include concerns about the appearance of the panels, the cost of installation, or uncertainty about the technology. A skilled solar energy salesperson should be prepared to address these concerns and provide clear and accurate information to help customers make an informed decision.

In addition to these strategies, a successful solar energy salesperson should also be knowledgeable about the latest solar technology, regulatory environment, and industry trends. This can help build confidence in the salesperson and the product being sold.

Ultimately, the goal of closing a sale is to help the customer understand the benefits of solar energy and feel confident in their decision to make a purchase. By providing accurate information, building trust and credibility, and offering incentives and financing options, a solar energy salesperson can help customers make the switch to solar energy and contribute to a more sustainable energy future.

Providing Excellent Customer Service And Follow-Up

When it comes to selling solar energy, it's not enough to just offer a great product. You also need to provide excellent customer service and follow-up to help close the sale. Customers want to know that they are making the right decision when investing in solar energy, and they need to trust the company they are working with. In this chapter, we'll discuss the importance of customer service and follow-up, and how it can help you close the sale.

Building Relationships And Trust

Building strong relationships with potential customers is essential in the solar energy industry. By establishing trust and credibility, you can help ease any concerns they may have and provide them with the information they need to make an informed decision. A friendly, knowledgeable sales team who can answer all their questions can make a huge difference. Ensure that your sales team is trained to handle any questions and provide a positive experience for the customer.

Incentives And Financing Options

Offering incentives and financing options can help make solar energy more accessible to potential customers. Incentives, such as tax credits and rebates, can help reduce the initial cost of solar energy, making it more affordable for customers. Financing options, such as lease and power purchase agreements, can also make it easier for customers to invest in solar energy without having to pay the full cost upfront.

Thesolarthinking.net

Excellent Customer Service And Follow-up

Providing excellent customer service and follow-up is crucial in closing the sale. After the initial meeting, ensure that your sales team follows up with potential customers to answer any remaining questions, provide additional information, and address any concerns. Follow-up can also help keep the customer engaged and interested in the product, leading to a higher likelihood of a sale. Make sure to be available and responsive to customer inquiries and concerns. If they have an issue, make it a priority to resolve it quickly and efficiently.

Staying Connected

After the sale, make sure to stay connected with your customers. Send them regular updates on their system's performance, offer maintenance and repair services, and provide them with any new products and services. Staying connected with customers can help build long-lasting relationships, leading to repeat business and referrals.

In conclusion, providing excellent customer service and follow-up is essential in closing the sale in solar energy marketing.

Building relationships and trust with potential customers, offering incentives and financing options, and staying connected with customers after the sale can make a significant impact in the success of your solar energy business. Remember, solar energy is not just a product, it's an investment, and customers need to trust the company they are working with to ensure the investment pays off.

Thesolarthinking.net

The Solar King's Advice to Solar Sales Professionals

Learning never stops when you are a sales professional. You may consistently close deals and even become the top sales rep in your company but you can never stop working to get better or you will lose your edge.

I have personally invested over $75,000 in training courses, masterminds, networking events, conferences, and mentorships but I still make it a point to continue to improve every day. Even after I became the top sales rep at my company I still continued to strive to get better.

I encourage you to do the same.

The Solar King's Additional Resources For Solar Marketers And Sales Organizations

- The Solar King's Free Solar Marketing Resources
  https://thesolarking.net/free-solar-resources
- The Solar King's Digital Marketing Course
  https://thesolarking.net/solar-leads-course
- The Solar King's Done For Your Marketing Package:
  https://thesolarking.net/offer
- The Solar King's Coaching Programs
  https://thesolarking.net/coaching
- The Solar King's Live Training Events
  https://thesolarking.net/events

Conclusion

To successfully market and sell solar energy, it's important to have a comprehensive approach that incorporates several key elements. These include understanding your target market, developing a marketing plan, utilizing both digital and traditional marketing techniques, overcoming challenges, closing the sale, and measuring success.

Understanding your target market involves getting to know their needs, preferences, and values to tailor your messaging and approach to best resonate with them. A well-developed marketing plan should include an analysis of your competition, a positioning statement, and a marketing mix strategy that outlines specific tactics to reach and engage your target audience.

Utilizing digital marketing, such as social media, email, and search engine optimization, can help expand your online presence and reach a wider audience. But don't overlook traditional marketing techniques, such as direct mail, print advertising, and radio or television ads, which can still be effective.

Selling solar energy can present unique challenges, including the high upfront costs of installations and consumer skepticism about the technology. Sales and marketing professionals must overcome these challenges through effective communication, education, and customer service.

To close the sale, sales professionals need to effectively communicate the benefits of solar energy and build a trusting relationship with customers. And measuring success, by tracking metrics like website traffic, lead generation, and sales conversions, helps refine your approach and identify areas for improvement.

By staying informed about industry trends, understanding your target market, and continually refining your approach, you can successfully market and sell solar energy, contributing to the growth of the renewable energy industry.

The solar energy industry is poised for significant growth and success in the coming years. As the world becomes more conscious of the environmental impact of traditional energy sources, there is an increasing demand for renewable energy solutions like solar power.

One of the key drivers of growth in the solar energy industry is the decreasing cost of solar panel installation. In recent years, the cost of solar panels has declined significantly, making solar energy a more accessible and cost-effective option for households and businesses.

Another factor contributing to the growth of the solar energy industry is the increasing availability of financing options. As more banks and financial institutions recognize the potential of solar energy, they are offering loans and other financing options to help homeowners and businesses finance solar panel installations.

Government policies and incentives also play a significant role in driving the growth of the solar energy industry. Many governments around the world offer tax credits, rebates, and other incentives to encourage the adoption of renewable energy solutions like solar power.

The growing popularity of electric vehicles is also contributing to the growth of the solar energy industry. As more households and businesses switch to electric vehicles, the demand for electricity will increase, making solar energy an even more attractive option for meeting this demand.

Additionally, the advancement of solar technology is helping to increase the efficiency and effectiveness of solar energy systems. New materials and designs are being developed to improve the efficiency of solar panels, making them more effective at generating electricity. Overall, the potential for growth and success in the solar energy industry is significant.

As the world becomes more conscious of the environmental impact of traditional energy sources, the demand for renewable energy solutions like solar power will only continue to increase. By leveraging new technology, government policies and incentives, and innovative financing options, the solar energy industry is poised to become a major player in the global energy market in the coming years.

The renewable energy industry is rapidly growing, and solar energy is at the forefront of this expansion. With an increasing number of households and businesses turning to solar power, it's a great time to pursue a career in the solar energy sector.

Solar energy offers a range of exciting and rewarding career opportunities, from sales and marketing to installation and maintenance. Not only does it provide job security and stability, but it also offers the chance to be a part of a vital industry that's helping to shape the future of energy.

To stay ahead of the competition and succeed in the solar energy industry, it's essential to keep learning and adapting. As with any rapidly growing industry, there are new developments and advancements all the time, and staying on top of the latest trends and technologies can give you a competitive edge.

One way to stay current is to pursue continuing education and training opportunities. Many colleges and universities now offer courses and programs focused on solar energy, providing a chance to gain specialized knowledge and skills. Additionally, professional organizations and trade associations provide access to industry events, conferences, and networking opportunities.

It's also important to be adaptable and open to change. The solar energy industry is constantly evolving, and being able to adjust to new trends and innovations is essential for staying competitive. Whether it's embracing new technologies, developing new marketing strategies, or adapting to changes in the regulatory environment, being flexible and responsive can help you stay ahead of the curve.

Overall, pursuing a career in solar energy is an exciting and rewarding opportunity, with the potential for personal and professional growth. By staying informed, continuing to learn and adapt, and embracing the opportunities presented by this dynamic industry, you can succeed and thrive in the solar energy sector.

The Solar King's Additional Resources For Solar Marketers And Sales Organizations

- The Solar King's Free Solar Marketing Resources
  https://thesolarking.net/free-solar-resources
- The Solar King's Digital Marketing Course
  https://thesolarking.net/solar-leads-course
- The Solar King's Done For Your Marketing Package:
  https://thesolarking.net/offer
- The Solar King's Coaching Programs
  https://thesolarking.net/coaching
- The Solar King's Live Training Events
  https://thesolarking.net/events

www.ingramcontent.com/pod-product-compliance
Lightning Source LLC
Chambersburg PA
CBHW060854170526
45158CB00001B/349